A Golden Thread

A Golden Thread

2500 Years of Solar Architecture and Technology

by Ken Butti and John Perlin

Cheshire Books, Palo Alto

Van Nostrand Reinhold Company, New York
Cincinnati, Toronto, London, Melbourne

Printed in the United States of America

Produced by Cheshire Books
514 Bryant Street, Palo Alto, CA 94301, U.S.A.

Published by Van Nostrand Reinhold Company
A division of Litton Educational Publishing, Inc.
135 West 50th Street, New York, NY 10020, U.S.A.

Van Nostrand Reinhold Limited
1410 Birchmount Road
Scarborough, Ontario M1P 2E7, Canada

Van Nostrand Reinhold Australia Pty. Ltd.
17 Queen Street
Mitcham, Victoria 3132, Australia

Van Nostrand Reinhold Company Limited
Molly Millars Lane
Wokingham, Berkshire, England

Library of Congress Catalog Card Number 79-25095
ISBN 0-442-24005-8 (Van Nostrand Reinhold)
ISBN 0-917352-07-6 (Cheshire Books)

16 15 14 13 12 11 10 9 8 7 6 5 4 3 2 1

Acknowledgments
Linda Goodman: Art direction, Cover design
Edward Wong-Ligda: Book design
Sara Boore: Line drawings
Gloria Jewel Leitner: Copy editor
Michael Riordan: Editor

Back Cover Photos, courtesy of:
M.I.T. Historical Collections (middle, left)
Walter Herdeg (middle, center)
Michael Vaccaro (middle, right)
National Aeronautics and Space Administration
(bottom, left)
Los Angeles County Museum of Natural History
(bottom, center)

Library of Congress Cataloging in Publication Data

Butti, Ken.
    A golden thread.

    Bibliography: p.
    Includes index.
    1. Solar energy—History. 2. Architecture and
solar radiation—History. I. Perlin, John, joint
author. II. Title.
TJ810.B88   621.47'09   79-25095
ISBN 0-442-24005-8

Authors' Note

Almost 2,500 years ago, the ancient Greeks began designing their homes to capture winter sunlight. This book describes the major advances in solar architecture and technology that have occurred since that time. Its emphasis is on developments occurring in Western civilization, but it also touches on important work from China and Japan. In writing this book, we took an evolutionary approach. We have tried to show the step-by-step development of solar architecture and technology, and to describe the process of discovery and the popularization of these applications. Thus, technological developments are always discussed within the economic, political, social and cultural milieux of each period.

In writing this history, we obtained help from many people. We would especially like to thank the following individuals who gave generously of their time and effort—*Greek and Roman Solar Architecture:* Dr. Borimir Jordan, Professor of Classics at the University of California at Santa Barbara; Dr. J. Walter Graham, Professor Emeritus of Fine Arts at the University of Toronto; and Dr. Norman Neuerburg, Professor of Art at California State University, Dominguez Hills. *Power from the Sun:* Frank Shuman. *Solar Water Heating:* William J. Bailey, Jr., Edward Arthur, William Crandall, Harold Heath, and Walter van Rossem. *Solar House Heating:* Wilhelm von Moltke, former Chairman of the Department of City and Regional Planning, Harvard Graduate School of Design: George Fred Keck; Henry (N.) Wright, former managing editor of *Architectural Forum*; Arthur Brown, A.I.A.; Dr. Albert G. H. Dietz, Professor of Architectural Engineering at the Massachusetts Institute of Technology; and Dr. George Löf, director of the Colorado State University solar research program. *The Modern Era:* Gonen Yissar; Professor Ichimatsu Tanashita, President of Ikutoku Technical University of Tokyo; Darryl M. Chapin; and Charles A. Scarlott. In addition, we appreciate the contributions of the many other people listed at the back of this book.

We thank Gloria Jewel Leitner and Dr. Michael Riordan for their assistance in preparing the manuscript. Gloria stood by us through several lean years, always believing in the worth of our project. Contributing valuable ideas—including the title of the book—and ferreting out contradictions in the text, she helped us convert our original manuscript into a much more readable form. Michael made sure our technical assertions were accurate. He also spent many hours poring over the manuscript, and his critical abilities contributed much to the book's final form.

Graphic design is no small small part of this book, and we thank Linda Goodman for helping us organize a vast amount of material into a coherent, striking whole. Edward Wong collaborated with her on book design, and Sara Boore provided a number of fine drawings.

Frank and Marie Hall and Luba and Paul Perlin generously helped with expenses in what appeared to be an endless project. The Irving F. Laucks Foundation of Santa Barbara also provided financial support. Edward Engberg helped guide us through the trying publishing world. We also thank Amory Lovins for his excellent Foreword and John Yellott, who reviewed the entire manuscript.

Portions of this book have already appeared in *Chicago Magazine, CoEvolution Quarterly, Solar Age, Solar Energy Handbook* (McGraw-Hill), and *Solar Law Reporter*. For the reader wishing to examine our evidence or to do further research in the history of solar energy, we have provided an extensive list of footnotes at the back of the book.

Ken Butti and John Perlin
October 1979

Table of Contents

While dining with some friends in Rome last year, I mentioned having just read in *Scientific American* a splendid analysis of ancient Persian buildings which had used sophisticated passive solar designs to stay cool in the midst of a blazing desert. I knew that similarly effective designs had been used in the Indian Pueblos of the southwestern United States and in ancient Chinese architecture to ensure coolness in summer and warmth in winter. To my astonishment, my dining companions proceeded to describe a coastal Italian town which had been fitted with central air conditioning thousands of years ago. A steady moist breeze off the Mediterranean blew against a cliff atop which stood the town. Upon entering cool caves in the face of the cliff, the breeze was chilled and dropped its condensed water vapor into ponds. The cool air was then taken up through shafts drilled down into the back of the caves and distributed throughout the city by stone ducts. And the distilled water in the ponds served as a pure and plentiful supply to the townspeople as well!

With this conversation fresh in my mind, it was with special delight that I learned that Ken Butti and John Perlin were writing *A Golden Thread*. The general public needs such a clear and evocative account of how solar architecture and technology have evolved in response to the social and economic forces of the past few thousand years. Even though it covers only highlights in the history of direct use of solar energy, *A Golden Thread* is rich with surprises on nearly every page. It will teach even the most erudite solar experts much about their own subject.

Our society abounds with people who be-

Foreword

lieve that solar energy is an exotic new source of energy requiring decades of further research and development before it can be proven practical. Butti and Perlin show that, on the contrary, solar power, water heating and home heating technologies have been evolving for thousands of years. They have been passed from one culture to another—developing into forms ever more suited to social needs and ever more ingenious in their use of new scientific knowledge and better materials. This improvement continues today at an unprecedented rate; the best present-day art in most solar technologies is much improved from that of a few decades or even a few years ago. But the effectiveness of many of even the oldest solar technologies, especially the simpler ones like "passive" solar architecture, has been adequate for centuries.

The steady evolution of solar architecture and technology has been periodically interrupted by the discovery of apparently plentiful and cheap fuels such as new forests or deposits of coal, oil, natural gas and uranium. Successive civilizations have shortsightedly treated this energy capital as income. Barbara Ward reminds us that like the Spanish Empire, eventually destroyed by the unsustainable wealth from a flood of New World gold, these civilizations have taken leave of their senses when confronted by short-term fuel supplies adequate to support a few generations' binge. This attitude persists today. We speak of "producing" oil, as if it were made in a factory; but only God produces oil, and all we know is how to mine it and burn it up. Neglecting the interests of future generations who are not here to bid on this oil, we have been squandering in the

last few decades a patrimony of hundreds of millions of years. Only recently have we begun to come full circle to the same realization that similar boom-and-bust cultures have reached before us: that we must turn back to the sun and seek elegant ways to live within the renewable energy income that it bestows on us.

The current energy "crisis" is nothing new, and it is very important to appreciate the lessons of earlier crises lest we repeat them. Earlier cultures—from the wood-short Greeks and Romans onward—became aware of the limits of their dwindling fuel resources. In the disastrous U.S. coal strikes around the end of the last century, the vulnerability of the nation to disruption in particular social groups and distribution networks became painfully apparent. These threatened cultures then rediscovered much of the earlier knowledge of permanent, practical solar energy. At several piquant moments in history—the latest of them today—wise observers of the energy scene have bemoaned the absurdity of having to rediscover and reinvent what should have been practiced continuously. They have been astonished to discover how much "novel" and "exotic" solar technology was in fact old and well understood.

Today, then, we stand precisely in the place several earlier cultures have stood. We have suddenly learned the transitory and ephemeral nature, the vulnerability, and the high social, ecological and even economic costs of depending on nonrenewable hydrocarbons to hold our societies together. But still we are playing elaborate games of self-deception: we give these precious fuels—and the electricity made from them—tax and price subsidies which in the U.S. totalled roughly a hundred billion dollars in 1978 alone. These subsidies are so lavish that they often outstrip such laudable efforts as California's 55 percent solar tax credit, and solar energy that is in reality considerably cheaper than Alaskan gas can end up looking more expensive. Further, we only require such "conventional" investments as different kinds of new power stations to compete with each other economically. But we require solar energy to compete with the heavily subsidized oil and gas which it is meant to replace. Our government therefore rejects as uneconomic—more expensive than oil at $15 a barrel—the more expensive kinds of solar heating and biomass liquid fuels at $20 to $25 a barrel. But it simultaneously seeks to extract from our pockets the most lavish and bizarre subsidies for synthetic gas at $30 to $40 a barrel or nuclear electricity at $100 per barrel equivalent. Of course that's crazy, but that's just what we're doing in a frantic attempt to divert public attention away from the economic reality of practical solar energy. Though some presently available solar technologies—not all—are somewhat more expensive than the old oil and gas, almost all cost several times *less* than what we would otherwise have to pay to replace them with nuclear power stations or synthetic fuels.

People who point out the obvious merits of solar energy still face the same obstacles they have always faced: lack of fair and symmetrical tests (how do the costs, rates and risks of solar energy compare with those of not having it?), misinformation from competing vested interests, and above all a widespread ignorance of what has actually been accomplished, both re-

cently and in prior ages. How many who today say that passive suncatchers cannot be added to heat urban buildings notice the lovely Victorian glassed arcades and conservatories throughout northern Europe? In 1977 and 1978, as the United States government was repeating the fiction that passive solar heating can only be used in new buildings specially designed for the purpose, over 25,000 American homeowners proved them wrong by adding south-facing greenhouses to their homes. How many who, like the authors of some Arizona building codes, doubt the durability and comfort of classical adobe buildings in the American Southwest have been inside an ancient Anasazi dwelling? How many wheels must we reinvent before we stop making them square?

Some of the technical lessons of solar energy found in *A Golden Thread* are still being relearned today. The drawbacks of high-temperature solar concentrators for driving heat engines were discovered almost a century ago. These same drawbacks are today leading many of our best analysts to turn from solar "power towers" to low-technology, low-temperature systems like solar ponds with Rankine cycle engines. Likewise, it is today the conventional wisdom to advocate extensive use of some of the more expensive solar technologies in remote rural areas and in the Third World—where conventional energy costs are prohibitively high. Exactly the same logic led to the pioneering early work in solar irrigation pumping in the American Southwest and in the French and British colonies of North Africa. Once again we have come full circle.

But perhaps this is the last time we shall have to do so. Perhaps this is the last time the inevitable solar age will be temporarily fore-stalled by a false sense of abundance sparked by a flash in the pan. For unless some form of energy now wholly unknown is discovered soon, there are no long-term energy alternatives other than nuclear reactions kindled artificially on earth and the natural energy flows driven by nuclear fusion remotely sited at the appropriate distance of 93 million miles.

In my opinion the choice has already been made. Because a soft energy path is the only one we can really afford—economically, politically and environmentally—it is inevitable, subject only to how difficult we want to make it for ourselves. Indeed, despite the obstacles still placed in its path, the solar transition that is already underway is starting to gain a breadth and speed that astonish some of its most ardent advocates. The reasons are obvious. Solar technologies require days, weeks or months to build rather than a decade. They can be diffused readily into a vast consumer market rather than requiring tedious "technology delivery" to a narrow, specialized utility market. Thus they are clearly faster than alternatives like nuclear power. The very diversity of these appropriate solar technologies enables a large number of slowly growing contributions to add up independently to a very rapid total growth.

But to me the most exciting reason for the speed of the solar advance in the past few years is that such "vernacular technologies," to use Ivan Illich's phrase, are accessible to practically everyone and do not require decades of research by large, impersonal organizations. They can instead tap the immense initiative and diverse ingenuity of our society. (For this same reason, the most interesting solar research tends to be done outside the official, government-sponsored research community and at a pace far outstripping the government's ability to find out what has been done.) Already, according to the latest on-the-spot estimates, half the houses in Nova Scotia have been weatherstripped and insulated in just one year; 40 percent of Vermont households installed efficient wood stoves in three years—entirely on private initiative; in the most solar-conscious communities in the United States, a quarter or more of the 1978 housing starts were passively solar heated; by early 1979 the U.S. had some 100,000-odd solar installations—twice as many as Florida had before cheap electricity interrupted the trend. These examples suggest that, given the opportunity and incentive, people can solve their own energy problems with extraordinary speed. The first and most difficult step is realizing that it can be done, and in creating that consciousness this book makes a lasting contribution.

Amory B. Lovins
London, England
April 28, 1979

I
Early Use of the Sun

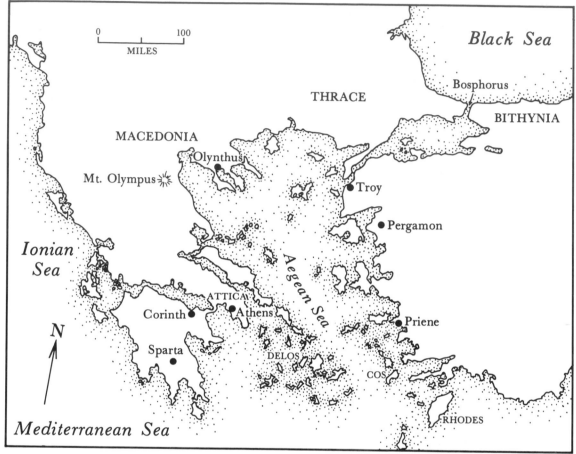

Map of Ancient Greece and Asia Minor.

Chapter 1

Solar Architecture in Ancient Greece

According to Socrates, the ideal house should be cool in summer and warm in winter. But Socrates' ideal was not easy to accomplish 2,500 years ago in ancient Greece. The Greeks had no artificial means of cooling their homes during the scorching summers; nor were their heating systems, mostly portable charcoal-burning braziers, adequate to keep them warm in winter.

Local fuel shortages probably exacerbated the problem. Near the populated areas people ravaged forests for wood to heat their homes and cook. Trees were also in demand to fuel smelting operations and build homes and ships. Goats foraging on saplings hastened the destruction of these timberlands. By the fifth century B.C., many parts of Greece were almost totally denuded of trees. Plato compared the hills and mountains of his native Attica to the bones of a wasted body: "All the richer and softer parts have fallen away," he lamented, "and the mere skeleton of the land remains."

As indigenous supplies dwindled and wood had to be imported, many city-states regulated the use of wood and charcoal. In the fourth century B.C., the Athenians banned the use of olive wood for making charcoal. Most probably they passed this measure to protect their valuable groves from incursions by fuel-hungry citizens. In the same century they also forbade the exportation of wood from nearby Attica. Yet Athens' own supply lines stretched all the way across Asia Minor to the shores of the Black Sea. On the island of Cos, the government taxed wood used for domestic heating and cooking. Authorities on Delos, which had no indigenous supplies whatever, severely restricted the sale of charcoal. They believed that such a valuable energy source should not be controlled by a few powerful merchants—leaving the consumers to pay any price.

With wood scarce and the sources of supply so far away, fuel prices most likely rose. Fortunately, an alternative source of energy was available—the sun—whose energy was plentiful and free. In many areas of Greece, the use of solar energy to help heat homes was a positive response to the energy shortage. Living in a climate that was sunny almost year-round, the Greeks learned to build their houses to take advantage of the sun's rays during the moderately cool winters, and to avoid the sun's heat during the hot summers. Thus solar architecture—designing buildings to make optimal use of the sun—was born in the West.

Modern excavations of many Classical Greek cities show that solar architecture flourished throughout the area. Individual homes were oriented toward the southern horizon, and entire cities were planned to allow their citizens equal access to the winter sun. A solar-oriented home allowed its inhabitants to depend less on charcoal-burning braziers—conserving fuel and saving money.

Because the Greeks venerated the sun, the development of solar architecture encountered few cultural impediments. Theophrastus, a noted naturalist of the time, commented that almost every citizen believed that—

> The sun provides the life-sustaining heat in animals and plants. It also probably supplies the heat of earthly flames. No doubt many people believe they are catching sun rays when making a fire.

It was also commonly held that exposure to the sun nurtured good health. Oribasius, an ancient medical authority, wrote that south-facing areas were healthy places because of their exposure to the sun. He also noted that north-facing areas

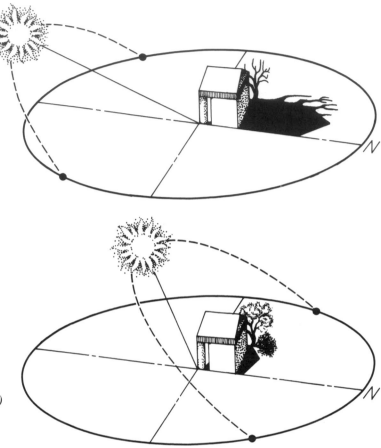

Path of the sun on December 21 (top) and June 21 (right), the winter and summer solstices. Solar position is shown for noon on both days.

were the least healthy, because they "do not receive much sun and when they do, the light falls obliquely without much vitality." With such a positive attitude toward the sun, the Greeks readily embraced the new solar building ideas of the time.

Principles of Solar Architecture

Solar architecture among the ancients was based on the changing position of the sun during different seasons. The Greeks' use of sundials made them keenly aware of daily and seasonal variations in the sun's course. The most common sundial of the period was the *gnomon*, a pole placed perpendicular to the ground. The Greeks could tell the hour, the day, and the season from the angle and length of the shadow cast by the sun as it traversed different parts of the sky during different times of the year.

The Greeks knew that in winter the sun travels in a low arc across the southern sky; in summer, it passes high overhead. They built their homes so that the winter sunlight could easily enter the house through a south-facing portico similar to a covered porch. There was no glass between the open-air portico and the entrances to the rooms inside the house, because the Greeks had neither clear glass nor other

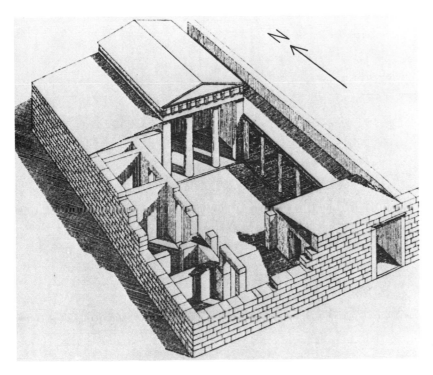

Reconstruction of a Classical Greek home, from excavations of the city of Priene by Theodore Wiegand. The rooms behind the portico faced south onto the courtyard.

transparent materials for doorways or windows. Not only were the main rooms in the house warmed by the rays of the sun streaming in through the portico, but they were sheltered from the north, as Aristotle noted, to keep out the cold winds. During the summer, overhanging eaves or roofs shaded the rooms of the house from the high sun during most of the day. As quoted by Xenophon, Socrates explained this approach:

> In houses that look toward the south, the sun penetrates the portico in winter, while in summer the path of the sun is right over our heads and above the roof so that there is shade.

These simple design principles served as the basis of solar architecture in ancient Greece.

Olynthus: A Planned Solar City

A new addition (North Hill) to Olynthus, one of the leading cities of northern Greece during Hellenic times, illustrates how the Greeks put solar architecture into practice in a densely populated community. The latitide of Olynthus was approximately that of New York and Chicago, and the temperature often dropped below freezing in winter. In the fifth century B.C., the yoke of Athenian domination became unbearable for the Olynthians and their neighbors living on the coast. They revolted and erected a more defensible perimeter against Athenian revenge by building a new community adjacent to the old city of Olynthus. Approximately 2,500 people lived there.

Right: Typical street plan in the North Hill section of Olynthus. This ancient community aligned its streets and avenues so that all buildings could face south.

Below: Socrates, an early advocate of solar architecture.

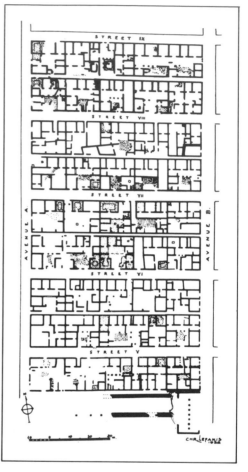

North Hill was a totally planned community—as were most urban centers built in Greece at the time. Starting from scratch, the settlers could more easily implement the new solar architecture of the day. They situated the town atop a sweeping plateau and built the streets perpendicular to each other, running north-south and east-west. In this way, all the houses on a street could be built with a southern exposure. This layout helped assure equality of housing for all residents—in keeping with the democratic ethos of the period. Aristotle commented that such rational planning was the "modern fashion;" it allowed a more convenient arrangement of homes to take maximum advantage of the sun.

Olynthian builders usually constructed a block-long row of houses simultaneously. The typical dwelling had six or more rooms on the ground floor and probably as many on the upper floor, averaging a total of 3,200 square feet of floor space. Each house was a standard square shape and shared a common foundation, roof and north wall with the other houses on the block. The north wall was made of adobe bricks approximately one and a half feet thick, which kept out the cold north winds of winter. If this wall had any window openings, they were few in number and were kept tightly shuttered during cold weather.

Top: Model of Olynthian apartments, southern face. Rows were built far enough apart to guarantee each home good exposure to the winter sun.

Above: Olynthian villa, scale model from the Royal Ontario Museum in Toronto. Overhangs commonly shaded the interior from the high summer sun.

The main living rooms of a house, each nearly 16 feet deep, faced a portico supported by wooden pillars located on the south side of the building. The portico led out to an open-air court averaging 320 square feet, which was separated from the street by a low wall. The court provided a place where the occupants could enjoy the out-of-doors with maximum privacy; it also served as the home's primary light and winter heat source.

In winter, rays from the sun traveling low across the southern sky streamed across the south-facing court, through the portico, and into the house—heating the main rooms. The earthen floors and adobe walls absorbed and retained much of this solar heat. In the evening, when the air inside began to cool, the floors and walls released this stored solar heat and helped warm the house. To prevent cold drafts from coming through the open portico into the house, some builders constructed a low adobe wall between the pillars of the portico, parallel to the south wall of the house. According to J. Walter Graham, one of the excavators of Olynthus, this strategy "allowed for the retention of the opening where it was most effective in

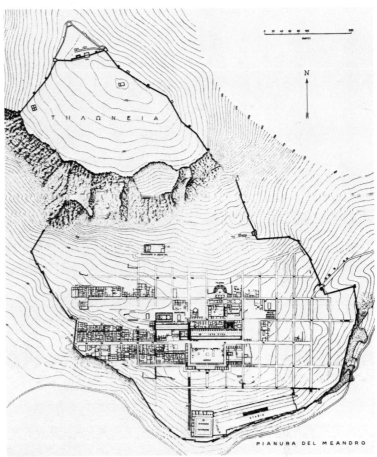

The ancient city of Priene (left), after relocation to the southern slope of Mount Mycale. Despite this difficult terrain, almost all the homes could capture the winter sun's warmth. The city plan (above) shows that its streets faced due south.

The ruins of Priene.

admitting the warmth of the sun in winter, while the wall below shut out the cold floor-level drafts.''

If a home had a second story, builders made sure that both the upper and lower main rooms had access to the winter sun. Socrates explained how this feat was accomplished: ''To obtain this result, the section of the house facing south must be built lower than the northern section in order not to cut off the winter sun.'' He also pointed out that the higher northern part of the house helped to block the cold north winds from the lower, southern part.

The Olynthian solar house design worked well in summer and winter. When the summer sun was almost directly overhead—from about ten in the morning until two in the afternoon—the portico shaded the main rooms of the house from the sun's harsh rays. In addition, the closed adobe walls and contiguous dwellings on the east and west sides protected the house from the morning and afternoon rays of the sun.

Priene and Delos

The varied terrain of Greece and her Asian colonies was not always as ideal as the flat plateau at Olynthus for building solar cities. Priene, located in Asia Minor, is a good example of how architects and planners coped with adverse topography.

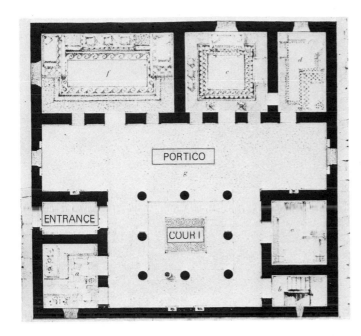

Floor plan of a house in Delos, from Exploration Archaelogique de Delos. *The main rooms at the back of this house, which closely resembled homes in the North Hill section of Olynthus, opened onto a covered porch and courtyard to the south (bottom of diagram).*

Plagued by constant flooding, the 4,000 residents of the old city of Priene decided to abandon their town for the safety of nearby Mount Mycale. The city planners devised a way to use the same Olynthian street plan on Mycale's steep slopes: the main avenues were terraced along the contours of the rocky spur on an east-west axis, while the secondary streets ran up the mountain from north to south. Due to the sharp incline, many of the secondary streets were actually stairways rather than roads.

Despite Priene's difficult location all homes, no matter how large or small, were designed according to what Priene's excavator, Theodore Wiegand, has called the solar building principle. The main rooms always led onto a south-facing, covered porch. Even homes belonging to the poorer citizenry could enjoy the warmth of the sun in winter and be spared its heat in summer.

Delos, an important trading center in the Aegean, presented even greater challenges to solar architects. The irregular rocky terrain of this island precluded the division of streets into an orderly pattern as at Olynthus or Priene. It also prevented uniform house plans. Often the contorted topography determined the plan of a Delian home. Nevertheless, the main rooms faced south whenever possible. Elsewhere, a new adaptation of solar design took form. Many residents terraced their homes along the sloping terrain so that the important rooms were on the upper level, with a commanding view of the south. If a southern orientation was impossible, a western exposure was the second choice. If these west-facing rooms became too hot in summer, the portico was closed off with curtains.

Solar architecture cut across class lines in ancient Greece. Rich and poor city dwellers as well as princes and kings relied on the sun. The important ceremonial and state rooms of the Palace at Pergamon faced south, as did the main rooms of commoners at Olynthus, Priene and, whenever possible, Delos. Many leading archaeologists, including J. Walter Graham and Theodore Wiegand, agree that solar

South facade (top) and interior (above) of a temple of ancient China. Winter sunlight streamed under the eaves and through latticework covered with silk or rice paper.

architecture was a prime concern of Classical Greek builders.

Solar architecture and urban planning evolved along similar lines in ancient China. The streets of important cities were always laid out in orthogonal fashion aligned with the points of the compass. Whenever the site permitted, the preferred house plan bore a striking resemblance to the Olynthian homes in that its principal apartments were built on the north side of a courtyard that opened to the south. While there were few window openings on the north, east and west walls, large wood-lattice windows covered by translucent rice paper or silk were common on the south. Decorative overhangs protected the interior from the hot summer sun.

How Well Did It Work?

According to Isomachus, a character in a Socratic dialogue by Xenophon, Greek techniques of solar architecture were quite effective. Isomachus brought his bride to his solar-oriented home and "showed her . . . living rooms for the family that are cool in summer and warm in winter." He told Socrates, "The whole house fronts south so that it is . . . sunny in winter and shady in summer."

Empirical evidence confirms Isomachus' praise. Edwin D. Thatcher, an architect, studied the solar heating capability of rooms facing south to determine the feasibility of indoor nude sunbathing during the winter. To simulate actual conditions, Thatcher relied on weather data for a climate similar to that of ancient Greece and Asia Minor. He found that a naked person sitting in the sunny part of such a room would be relatively comfortable on 67 percent of the days during the colder months of November through March. The room used for this study was not as well protected as an average Greek living room—and of course the residents would have been clothed most of the time. It seems safe to say that for most of the winter the sun would adequately heat the main rooms of a Greek solar-oriented home during the daytime. When solar heat was insufficient, charcoal braziers could be lit.

The great playwright Aeschylus suggested that a south-facing orientation was a normal characteristic of Greek homes. It was a sign of a "modern" or "civilized" dwelling, he declared, as opposed to houses built by primitives and barbarians who,

> Though they had eyes to see, they saw to no avail; they had ears, but understood not. But like shapes in dreams, throughout their time, without purpose they wrought all things in confusion. They lacked knowledge of houses . . . turned to face the sun, dwelling beneath the ground like swarming ants in sunless caves.

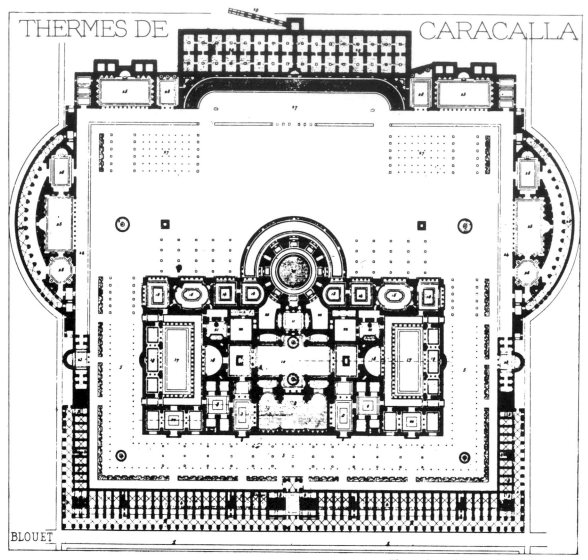

Floor plan of the Baths of Caracalla.

Chapter 2
Roman Solar Architecture

Wood consumption in ancient Rome was even more profligate than in Greece. There was a heavy demand for wood as fuel for industry, to build houses and ships, and to heat public baths and private villas. By the time of Christ, it was common for wealthy Romans to have central heating in their expansive villas. Their *hypocausts* burned wood or charcoal in furnaces and circulated the hot air through hollow bricks in the floors and walls. A *hypocaust* system could devour as much as 280 pounds of wood per hour, or more than two cords a day.

Indigenous wood supplies quickly disappeared from the Italian peninsula. The Mount Cimino region, a short distance from Rome, had been so heavily forested in the third century B.C. that the ancient historian Livy called it "more impassable and terrifying than . . . the wooded regions of Germany." But by the first century B.C., timber had to be imported from as far east as the Caucausus, more than a thousand miles away. A century later the geographer Strabo reported that a lack of fuel forced the islanders of Elba to shut down their iron mines. The natural philosopher Pliny the Elder described the adverse effects of wood shortages in the Campania region upon the local metal industry.

These local fuel shortages and the high cost of imported wood probably influenced the Romans to adopt Greek techniques of solar architecture. The Romans did more than copy the Greeks, however; they advanced solar technology by adapting home building design to different climates, using clear window coverings such as glass to enhance the effectiveness of solar heating, and expanding solar architecture to include greenhouses and huge public bath houses. Solar architecture became so much a part of Roman life that sun-rights guarantees were eventually enacted into Roman law.

Solar Design in Different Climates

Because the Roman Empire covered a far greater area than the Greek city-states, its architecture had to be tailored to many different environments. Vitruvius, the preeminent Roman architect of the first century B.C., counseled:

> We must begin by taking note of the countries and climates in which homes are to be built if our designs for them are to be correct. One type of house seems appropriate for Egypt, another for Spain . . . one still different for Rome, and so on with lands and countries of varying characteristics. This is because one part of the Earth is directly under the sun's course, another is far away from it, while another lies midway between these two. . . . It is obvious that designs for homes ought to conform to diversities of climate.

Accordingly, Vitruvius recommended that in extremely hot regions such as North Africa, where it was desirable to avoid solar heat, buildings should be open to the north—away from the sun. In more temperate parts of the Empire, such as the Italian peninsula, "buildings should be thoroughly shut in rather than exposed toward the north, and the main portion should face the warmer [south] side."

Because he was a student of Greek architecture, Vitruvius echoed the earlier advice of Socrates and Aristotle. But he went beyond his Greek mentors and specified where certain rooms should be placed for optimum comfort. He recommended that Romans living in temperate climates make sure their winter dining

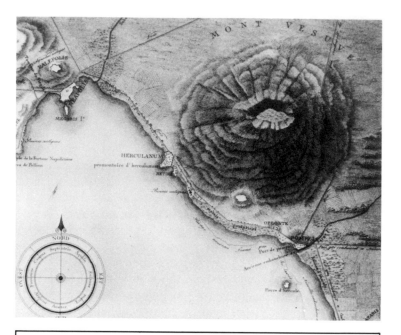

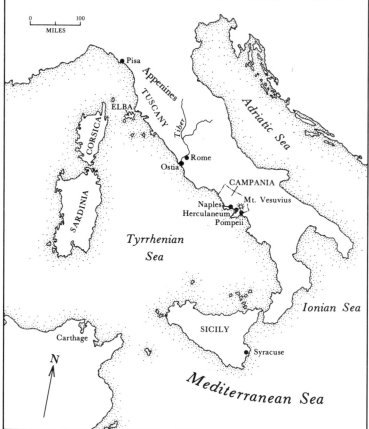

Right: Map of Italy during the first century A.D. Detail (top) shows the vicinity of Herculaneum and Pompeii, near Mount Vesuvius.

One of the large villas at Herculaneum. The sun rooms faced south and had large windows probably covered with glass or mica.

rooms looked to the winter sunset "because when the setting sun faces us with all its splendor, it gives off heat and renders this area warmer in the evening." But summer dining rooms "should have a northern aspect. For while the other aspects, at the solstice, are rendered oppressive by the heat, the northern aspect, because it is turned away from the sun's course, is always cool."

Rome learned about solar building techniques through the writings of Vitruvius, and also from direct contact with the Greek colonists who had settled throughout southern Italy. These towns succumbed to Rome as she extended her power down the Italian peninsula and into Sicily. One such city was Herculaneum near Mount Vesuvius, subdued by Sulla in 89 B.C. Soon after Herculaneum's fall, the dynamic Roman spirit took hold. Old homes were separated into apartments and office buildings, and new housing was built. While the growing masses of urban poor had to live wherever they could, the rich chose their own home sites and built according to the principles of good solar design. As at Olynthus and Priene, the main rooms of these new homes almost always opened to the south.

Pliny's Two Villas

As timber supplies dwindled during the first century A.D., wealthy Romans throughout the Empire began to build with the sun in mind. Those rich enough to own a country villa had fewer building restrictions than city dwellers and could employ solar design more easily. Pliny the Younger, a wealthy and influential writer of second-century Rome, described in letters to his friends how he used the sun to help heat his two villas. Pliny's motive was apparently twofold—to conserve wood and to save money. Although nearby forests provided sufficient firewood, Pliny was probably concerned about their eventual depletion. Once this source was consumed, where could he turn? Pliny also had an eye on his wallet. As he wrote to his friend Gallus, he wanted his villas large enough for his convenience without being expensive to maintain. A solar-oriented home would require a smaller furnace and

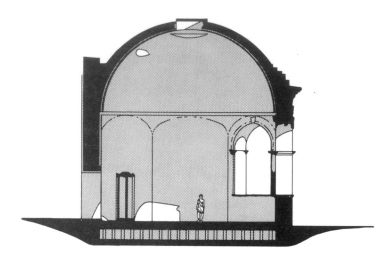

Top right: Cross-section of a Roman heliocaminus *in the public baths at Ostia, from studies by Edwin D. Thatcher. In winter, the large, transparent windows admitted plentiful sunlight but slowed the escape of solar heat.*

Right: Ruins of the heliocaminus *in Hadrian's villa at Tivoli. Large windows covered with glass or transparent stone faced south, southeast and southwest.*

fewer heating ducts than a similar non-solar, centrally heated villa that consumed more fuel.

Pliny's summer house stood in the foothills of the Appenines in north central Italy, where the summers remained fairly cool and the winters were severe. The main part of this villa was exposed to the south, allowing the sun "to enter by midday in summer, but much earlier in winter." Pliny built his winter retreat at

Laurentum, on the seacoast—close enough to Rome so that he could occasionally spend some time there after finishing work in the city. The dining room faced southwest, as Vitruvius had advocated, so that the room would be heated by rays of the setting winter sun and thereby provide a congenial atmosphere in which to dine. Southeast of the dining room was a large bedroom and beyond it a smaller one whose windows admitted both the morning and evening sun. The study, where Pliny spent much of the day reading, was semicircular with a large bay window that let in sunlight from morning to evening.

Glass As A Solar Heat Trap

Pliny called one of his favorite rooms a *heliocaminus*—literally "solar furnace"! Such a name indicates that this room got much hotter than the other rooms of his villa. Most probably the southwest openings of the *heliocaminus* were covered with glass or mica—a thin sheet of transparent stone. Such materials act as solar heat traps by admitting sunlight into a room and holding in the heat that accumulates inside. The solar-heated air inside such a room could not readily escape through these window coverings; so the temperature inside a *heliocaminus* rose well above what was possible in a Greek-style solar-oriented home without glazed window openings.

Pliny could easily have purchased window coverings for his *heliocaminus*, since a thriving window industry existed in Rome at the time. Transparent coverings made of thin stone such as mica or selenite were produced by splitting the stone into plates as thin as desired. Glass windows were probably fabricated by pouring molten glass into molds or onto a flat surface and pressing it with a roller. Another process involved blowing hot glass into a bubble, swinging it until it formed a cylindrical shape, and cutting the cylinder with iron shears before flattening out the pieces on an iron plate covered with sand.

Windows of transparent stone or glass were a radical innovation. Colored glass had been used for almost 3,000 years. But it was not until sometime in the first century A.D. that an anonymous Roman thought of using transparent materials to make windows that would let in light but keep out rain, snow and cold. The philosopher Seneca noted the newness of this idea in a letter written in A.D. 65, "Certain inventions have come about within our own memory—the use of window panes which admit light through a transparent material, for example."

Excavations at Herculaneum, Pompeii and elsewhere provide evidence that glass was occasionally made into windows for the homes of wealthy Romans—even in the early days of the Empire. Pliny the Elder observed in the first century that these new window panes, unlike opaque window coverings, allowed sunlight to enter a building. By the following century, the use of glass and other transparent materials to cover windows was increasingly common.

The Romans also used this solar heat trap principle to cultivate plants. Greenhouses were built to keep plants warm so that they would mature more quickly and produce fruits and vegetables even in winter. Emperor Tiberius, for example, had a penchant for cucumbers year-round. As Pliny the Elder noted, "there was never a day in which he went without." To oblige the ruler's appetite, the kitchen gardeners had cucumber beds mounted on wheels which they moved

Engraving (right) and photo (below) of "old style" Roman baths, from the first century B.C. Only a small chink in the ceiling admitted sunlight.

into the sun. During the colder days of winter, these gardeners placed "cold frames" glazed with transparent material over the plant beds to hold in the solar heat.

Wealthy Romans used transparent coverings to keep their exotic plants healthy during inclement weather. The favorite plants of the rich often basked in more warmth than did the poorer Roman citizens. Martial, the first-century satirist, could not resist parodying this inequity. He complained that while his patron protected exotic trees from the cold by placing them in a structure with a glazed south-facing wall that caught the sun's rays, the broken windows of his own room went unrepaired. "Not even the chilly winds will stay in my room," the poet woefully complained to his patron. "Do you wish your old friend to stay until he freezes? I should be better off as the guest of your trees!"

The Roman Baths

Another Roman application of the solar heat trap principle was the use of transparent window coverings to help heat their public baths. Probably no other people in history have cherished baths as much as they. From the first century A.D. onward, the public baths became immensely popular gathering places. People from all walks of life congregated there, usually in the late afternoon when the workday was over—to exercise in the gymnasium, take dips in the cold, warm and hot baths, sweat in the steam room, or lounge around listening to music, poetry and gossip.

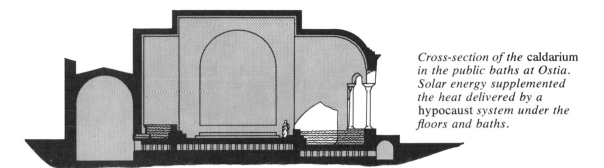

Cross-section of the caldarium *in the public baths at Ostia. Solar energy supplemented the heat delivered by a* hypocaust *system under the floors and baths.*

The baths of Caracalla, one of the largest Roman baths ever built, held as many as 2,000 people at a time. Seneca took a somewhat cynical view of the tumult inside these baths on a typical busy afternoon:

> Picture to yourself the assortment of sounds, which are strange enough to make me hate my very powers of hearing! When your strenuous gentleman, for example, is exercising himself by flourishing leaden weights; when he is working hard, or else pretends to be working hard, I can hear him grunt; and whenever he releases his imprisoned breath I can hear him panting in wheezy and high-pitched tones. Or perhaps I notice some lazy fellow, content with a cheap rub-down, and the crack of the pummeling hand on his shoulder varies in sound as the hand is laid on flat or hollow. . . . Add to this the arresting of an . . . occasional pickpocket, the racket of the man who always likes to hear his own voice in the bath, or the enthusiast who plunges into the swimming pool with unconscionable noise and splashing. . . . Then the cake seller with his different cries, the sausageman, the confectioner, and all the vendors of food hawking their wares, each with his own distinctive intonation.

The Roman baths were not always so boisterous. When public baths were first introduced in the second century B.C., they were small, modest establishments serving primarily as a place for washing. These baths were as dark as caverns. As Seneca remarked, they had "only tiny chinks—you cannot call them windows—cut out of stone" to let in air and light.

Two hundred years later, baths of great opulence sprang up throughout the Empire, including France and Britain. This was the age of Augustus; Rome now ruled most of Europe and large parts of North Africa, the Middle East and Asia Minor. Her citizens could afford to enjoy the spoils of conquest. Both public and private baths glittered with large and costly mirrors and mosaics, and Thracian marble lined the pools.

Despite such extravagance, the Romans displayed a utilitarian bent during this and later periods. As Vitruvius advocated:

> The site for the baths must be as warm as possible and turned away from the north. . . . They should look toward the winter sunset because when the setting sun faces us with all its splendor, it radiates heat, rendering this aspect warmer in the late afternoon.

From the first century A.D. onward, the builders of most bathing establishments followed this dictum. In almost every important bath house of the time, at least the

The Baths of Diocletian, from an early nineteenth-century engraving by Edmund Paulin. On a typical afternoon, thousands of Romans would be bathing, exercising and cavorting inside.

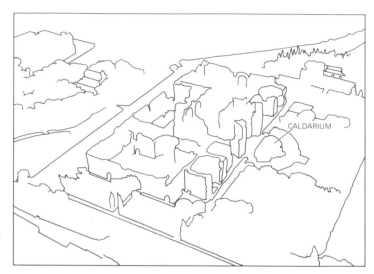

The ruins of the Baths of Caracalla in Rome. The hot bath or caldarium *was located in the semicircular area at right, facing southwest.*

hot bath, or *caldarium*, faced toward the winter sunset to absorb as much solar heat as possible while the majority of its users were cavorting inside. In addition, the Romans usually glazed the whole south wall of their bath houses. Seneca wrote that these giant windows trapped so much solar heat that by the late afternoon, bathers would "broil" inside the baths. In some of the more elaborate bath houses, the sweat room (which was usually semicircular in shape) boasted enormous windows

The south windows in the Central Baths at Pompeii stood almost ten feet high.

looking to the south and southwest, and also a sand floor, which absorbed the solar heat during the daytime and released it in the evening for the pleasure of late visitors.

Among the ruins of Pompeii, as in most Roman cities, there are examples of both the old-style Roman baths—structures "buried in darkness," as Seneca described them—and the later baths built to take advantage of the sun. According to August Mau, one of the excavators of Pompeii, there were only "small apertures, high in the walls and ceilings, through which light was admitted in the older baths." In contrast, the *caldarium* of the more "modern" Central Baths had south-facing windows measuring 6 ft 7 in. by 9 ft 10 in. to take maximum advantage of the afternoon sunlight.

The Romans preferred solar architecture for their bath houses because solar heat was thought to be healthier than artificial heat. Doctors considered the sun therapeutic for many ailments, from poor appetite to problems with "bodily fluid." But most important was the role of solar design in helping to cut down on the tremendous amount of fuel required to heat the bath water, steam room, and the air inside the building.

Rural Self-Reliance

By the fourth century A.D., the fuel situation had worsened. To satisfy the Roman fuel needs, the government commissioned an entire fleet of ships, called the *naviculari lignarii* (literally, "wood ships"), for the sole purpose of making wood runs from France and North Africa to Ostia, a Roman port. This suggests that almost all of the forests on the Italian peninsula had been ravaged by this time. Or perhaps Rome, now militarily on the defensive, could not keep her overland supply lines open at all times.

As border skirmishes with barbarians became more serious, Rome had to build up her military forces, requiring increased taxation and currency devaluations. Primarily these measures hurt her productive citizens. Rome's position grew more precarious despite her stepped-up military efforts, and the flow of vital goods into Rome from outlying areas of the Empire was disrupted. With a broken, dis-

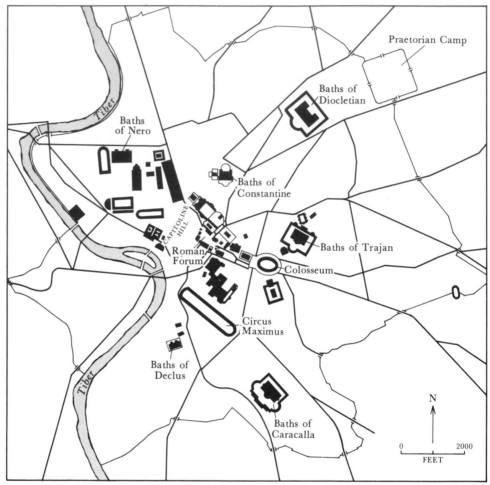

Map of Imperial Rome, showing locations of the major baths. Almost all the later, "new-style" baths faced south or southwest.

enchanted middle class and a scarcity of goods, Rome's urban economy fell into disarray. More and more wealthy citizens who owned land in the country left their city homes and settled on their estates. Cut off from the rest of the world, they were forced to adopt a self-reliant lifestyle.

To help the rich cope with their new rural way of life, Faventinus and later Palladius—the leading architects of the age—wrote building manuals stressing self-sufficiency. Both architects reiterated Vitruvius' advice on proper positioning of hot baths and living rooms, but they also offered energy-saving techniques of their own. Palladius advocated the recycling of bath water and the placement of winter rooms directly above the hot baths so that they would benefit from both the sun's heat and the waste heat rising from the baths below. And while Vitruvius had noted that olive oil storage rooms should face south to keep the oil from congealing in winter, Palladius added that a transparent window covering would give further protection.

Faventinus and Palladius recommended an ingenious way to make the floor of a sun-heated winter dining room an ideal absorber of solar energy. The technique had been invented earlier by the Greeks and passed on in the writings of Vitruvius. A shallow pit was to be dug under the floor and filled with broken earthenware or other rubble. On top a mixture of dark sand, ashes and lime was spread to form a black floor covering to serve as an excellent absorber of solar heat—especially during the afternoon. The mass of rubble underneath stored large amounts of heat and released it later in the evening when the room temperature cooled. Faventinus assured villa owners that such floors would stay warm during the dining hour and ''will please your servants, even those who go barefoot.''

Sun-Rights Laws

From the days of Augustus in the first century A.D. until the Fall of Rome, the use of solar energy to heat residences, bathing areas and greenhouses was apparently fairly widespread throughout the Empire. Exactly how much the Romans relied on the sun is impossible to say. But the *heliocaminus* or ''solar furnace'' room was common enough to provoke disputes over sun rights requiring adjudication in the Roman courts. As the population increased, buildings and other objects blocked the solar access of some *heliocamini*, and their owners sued. Ulpian, a jurist of the second century A.D., ruled in favor of the owners, declaring that a *heliocaminus'* access to sunshine could not be violated. His judgement was incorporated into the great Justinian Code of Law four centuries later:

> If any object is so placed as to take away the sunshine from a *heliocaminus*, it must be affirmed that this object creates a shadow in a place where sunshine is an absolute necessity. Thus it is in violation of the *heliocaminus'* right to the sun.

That this opinion was written into the Justinian Code in the sixth century strongly indicates that the construction of solar heated buildings continued until this late date.

Except for the community bath houses visited by the general public, solar heating was enjoyed by only the very rich in Rome. Mass use never caught on as it did in Greece. Rome did not plan for her poorer citizens. In sharp contrast to the Greek spirit of democracy and social equality, the dominant Roman ideology favored class privilege—so that only the wealthy could build their homes with the proper solar orientation. Technologically, however, Rome made important advances in solar architecture—namely, transparent window coverings used as solar heat traps for homes, baths and greenhouses. And for the first time in recorded history, laws were set down establishing sun rights.

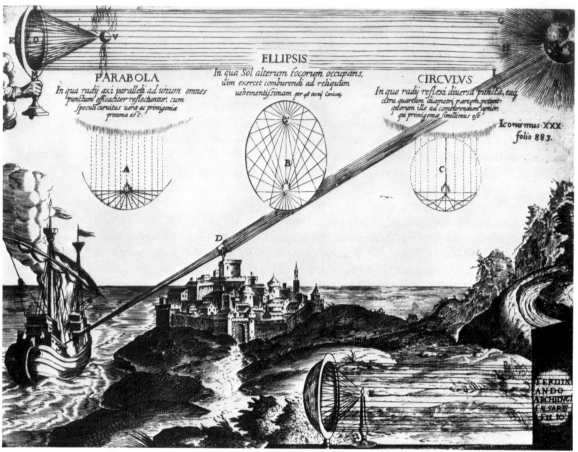

Engraving from Ars Magna Lucis et Umbrae, *by Athanasius Kircher, 1646. Because parabolic mirrors focus parallel rays of sunlight to a point, they can generate extremely high temperatures and ignite objects at a distance.*

Chapter 3

Burning Mirrors

The ancients knew of other ways to harness solar energy besides the solar orientation of buildings and use of glass as a solar heat trap. The Greeks, Romans and Chinese developed curved mirrors that could concentrate the sun's rays onto an object with enough intensity to make it burst into flames within seconds. Solar mirrors that could generate spectacular heat captured people's imaginations for centuries to come; they could be used constructively or possibly as terrible weapons of war.

The Greeks were the first Western people to describe "burning mirrors"—solar reflectors made of polished silver, copper or brass. The earliest burning mirrors probably consisted of many flat pieces of polished metal joined together to form a roughly curved surface. These were replaced by concave spherical mirrors made from single sheets of metal. Simpler in design, these spherical mirrors dispensed with the bulkiness of many small flat-plane mirrors linked together.

As their understanding of geometry matured, the Greeks realized that a mirror with a parabolic surface was even more powerful than a spherical mirror. Dositheius, a mathematician of the third century B.C., discovered that solar rays bouncing off a parabolic mirror are focussed almost to a point. Because the solar energy is concentrated into a smaller area than with a spherical mirror of similar size, a parabolic mirror produced higher temperatures. According to his contemporaries, Dositheius built the first parabolic mirror. A century later, in his treatise *On the Burning Mirror*, Diocles gave the first formal geometric proof of the focal properties of parabolic and spherical mirrors.

Legend has it that in 212 B.C. Archimedes used burning mirrors to destroy the ships of invading Romans at Syracuse, but this story appears to be no more than a myth. None of the reliable historians of the period—Polybius, Livy or Plutarch—mention that Archimedes used burning mirrors. However, his arsenal reportedly included other exotic weapons such as machines that dashed enemy craft to splinters with heavy beams, and crane-like devices that lifted boats into the air and smashed them against the rocky coast.

Burning mirrors did serve as a means of igniting fires in Greek and Roman times, but fires with more benign functions, such as those that burned in temples of worship. Plutarch wrote that when barbarians sacked the Temple of the Vestal Virgins at Delphi and extinguished their sacred flame, it had to be relit with the "pure and unpolluted flame from the sun." With "concave vessels of brass" the holy women directed the rays of the sun onto "light and dry matter" that was immediately ignited, and their flame burned anew.

The ancient Chinese also used burning mirrors primarily for religious purposes. Independently, they began making concave solar reflectors at about the same time as the Greeks. According to the *Chou Li*, a book of ceremonies written in approximately 20 A.D. which describes rituals dating far back into Chinese antiquity, "The Directors of Sun Fire have the duty of receiving, with a concave mirror, brilliant fire from the sun . . . in order to prepare brilliant torches for sacrifice."

Dreams of Solar Weaponry

As with so many achievements of antiquity, all knowledge of burning mirrors— their powers, methods of construction, and uses—vanished from European culture

Roger Bacon, an English scholar of the thirteenth century who urged the use of large burning mirrors in the Crusades. Bacon was eventually jailed for his proposals.

during the Dark Ages. Fortunately, the Arabs had a great reverence for learning during these times. They preserved, translated and elaborated upon many Classical works, including Greek geometry texts that discussed the mathematical properties of parabolic mirrors.

Ibn Al-Haitham was an eleventh-century Arab scholar living in Cairo, who had many of these ancient works at his disposal. The geometric proofs of Diocles were probably among them. Al-Haitham called burning mirrors "one of the noblest things conceived by geometricians of ancient times," but he felt that the Greeks did not convincingly explain their proofs. "Since in this matter there is great benefit and general usefulness," he wrote, "[I] have decided to explain and clarify it, so that those who seek truth will know the facts." Al-Haitham's elaborate mathematical proof was translated into Latin and circulated among several European universities during the middle of the thirteenth century. Consequently, his writing served as a bridge between scholars of medieval Europe and the ancient Greeks.

One of those privy to Al-Haitham's work was Roger Bacon, a Franciscan monk who taught at Oxford and the University of Paris during the thirteenth century. To Bacon, this essay on the concentrating powers of parabolic mirrors was more than idle academic chatter. It was a clue to the means of building a doomsday weapon that might be wielded by the Anti-Christ—the Moslems whom the Crusaders were battling in the Holy Land. Bacon advised Pope Clement IV that—

> This mirror would burn fiercely everything on which it could be focussed. We are to believe that the Anti-Christ will use these mirrors to burn up cities, camps and weapons.

He warned that an enemy Saracen (Al-Haitham) "shows in a book on burning mirrors how this instrument is made." But the specific details of its design were

Frontispiece from Al-Haitham's Opticae Thesaurus. *The use of burning mirrors for military aims was a favorite theme of Renaissance scholars.*

contained in another volume unavailable to Christendom—suggesting to Bacon sly deception on the part of the Arabs. Nevertheless, Bacon reassured the Pope, there was no need to worry: "The most skillful of Latins is busily engaged in the construction of this mirror." Whether this "Latin" was a colleague, as suggested by Bacon, or merely an indirect reference to himself is not known.

Several years later Bacon informed the Pope that his "colleague" had finally finished building a powerful parabolic mirror after working "many years . . . at great expense and labor . . . abandoning his studies and other necessary business." Certainly, if the Christians "living in the Holy Land had twelve such mirrors," Bacon advised the Pope, they could "expel the Saracens from their territory, avoiding any casualties on their side" and making it unnecessary for more Crusaders to intervene in the Middle East.

Whether Bacon ever really built this solar weapon or was just giving free rein to his imagination, we can only speculate. For such a device to be effective beyond the range of spears, arrows, slings, catapults and other weapons at the hands of enemy troops, it would have to be truly colossal—an unlikely achievement with the technology of the times. The mirror would also have to be moved throughout the day to stay in alignment with the sun. Such a feat would require an army, since a mirror of this size would be enormously heavy. There were other drawbacks to Bacon's scheme. The Saracens knew about solar reflectors themselves. Hence, they would be most likely to attack facing the sun, making it impossible for the mirror to reflect the sun's rays in their direction. Furthermore, burning mirrors can only work on sunny days. Only the parallel rays of direct sunlight can be concentrated on a small target area, and clouds tend to scatter solar rays in all directions. If the Mohammedans moved against the Christians on an overcast day—or at night, for that matter—Bacon's solar weapon would have been useless.

In his zeal, Bacon never informed the Pope of these limitations. However, this was the first time in medieval European history that someone had advocated the use of empirical science instead of metaphysical speculation. To make such a radical idea palatable, Bacon tried to show how technology could serve the interests of both princes and clerics. But his offer fell on deaf ears. Even though Pope Clement was liberal enough to tolerate the views of a proponent of experimental science, he never read any of Bacon's plans. Once the Pope died, a conservative wind swept through the Church. Empiricism threatened the entire world-view of most Christians—a view based on the metaphysics of Aristotle and divine revelation. The idea of transforming mild sun rays into a fierce weapon of fire was now condemned as the work of the devil, and Bacon was thrown into the dungeon as a heretic instead of receiving the funds to build the ultimate solar weapon that would help defend Christianity.

A Renaissance Revival

For several centuries after Bacon's suppression, natural philosophers were understandably less vocal about solar mirrors. Not until the sixteenth century was another fabulous mirror proposed, this time by Leonardo da Vinci. Its purpose would not be military, but peaceful—to generate heat and power for industry and recreation. Leonardo proposed to build a parabolic mirror four miles across that could "supply heat for any boiler in a dyeing factory, and with this a pool can be warmed up, because there will always be boiling water." Why the mirror had to be so enormous is not clear from his notes, nor is there any discussion of using many smaller mirrors in tandem to achieve the same end.

 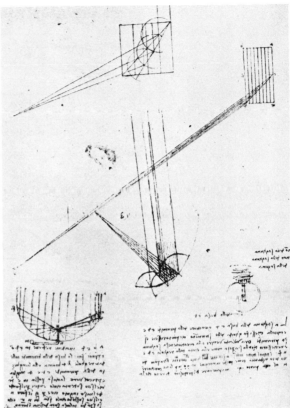

Left: Self-portrait of Leonardo da Vinci. Among his many achievements were the earliest known plans for industrial applications of solar energy.

Right: Excerpts from Leonardo's personal workbooks. Apparently he considered several designs for an enormous parabolic mirror.

Leonardo began building his giant mirror around 1515. Unlike the flamboyant Bacon, the Florentine kept his intentions a closely guarded secret. He purposely gave a misleading title—''Perspective''—to the folios containing his research on the power of concave mirrors, and jotted down the designs in a generally incomprehensible cipher. In spite of the personal importance of this project, Leonardo never completed his mirror, and the construction plans could not be decoded by others during his lifetime.

Other Renaissance artists and scientists had to be content with more modest projects using much smaller burning mirrors. According to Leonardo, the sculptor Andrea del Verrocchio employed a burning mirror to solder the sections of a copper ball lantern holder for the Santa Maria del Fiore Cathedral in Florence. Others used mirrors to make perfume. Adam Lonicier, writing in 1561, recorded the technique: alchemists submerged certain types of flowers in a water-filled vase, which was placed at the focal point of a spherical mirror; the concentrated solar heat caused the essence of the flowers to diffuse into the water.

With the aid of a burning mirror, sixteenth-century alchemists could manufacture perfume.

In Search of the Giant Mirror

During the latter part of the sixteenth and throughout the seventeenth century, almost every scientist and gentleman experimenter investigated the curious powers of burning mirrors. Giovanni Magini, the Italian astronomer, wrote that he could melt "lead, silver or gold in small quantities such as a coin held firmly with tongs." Magini's mirror, like most of the reflectors built at the time, was spherical and not parabolic. It was also quite modest in size—less than two feet in diameter.

Methods of production limited the size of burning mirrors built in the late 1500's and early 1600's. Most mirrors were fabricated from a highly lustrous alloy of copper, tin and arsenic. Craftsmen melted the alloy into a liquid, poured it into a concave mold, and then pried the mirror out after it had cooled and taken shape. The last step would have been practically impossible if the mirror were large, because the hardened metal would have been very heavy. A mirror measuring only two and a half feet in diameter weighed more than 110 pounds. Imagine the difficulty of lifting a mirror ten times this size! Moreover, the alloy was brittle, making it nearly impossible to lift a huge mirror out of its mold in one piece. Even if an alloy mirror of tremendous size could have been made, the same problem that Bacon ignored would still exist: an immense and heavy reflector would require enormous manpower to keep it turned toward the sun.

Still, mirror enthusiasts could dream. If a small device could produce so much heat, imagine the power of a mirror 100 times its size! Unfettered by strict empiri-

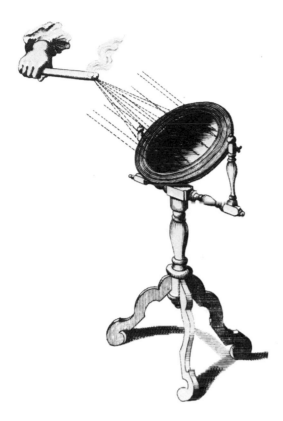

A small burning mirror typical of those used by natural scientists during the 1500's and 1600's.

cism, the experimenters of the age presented the theoretically plausible as established fact. Even a scientist of Galileo's caliber was fooled by these claims. He confessed that after "seeing a small mirror melt lead and burn every combustible substance, such effects as these render credible to me the marvels accomplished by the mirrors of Archimedes." The legend of this Greek's military feats with burning mirrors found a ready audience during this period.

Galileo's less rigorous contemporaries speculated even more wildly. Jerome Cardano was one of the most prolific encyclopedists of the time, and his volumes covered such varied topics as the debauchery of Caesar and the divisions of the animal kingdom. But Cardano did not rely on hard data when he boasted that he could build a mirror two miles in diameter with the capability of burning objects a mile away.

"Impossible!" exclaimed Giambattista Porta, a contemporary of Cardano, in a moment of healthy skepticism, " 'Tis strange how many follies he betrays himself guilty of." Porta elaborated further:

> To describe a circle whose diameter should be two miles long, what compass must we use and what plate shall we make it on, or who shall draw it about? The cause of his error was that he never made any such glass [mirror], for had he tried he would have spoken otherwise.

Unfortunately, Porta's defense of empiricism was fleeting—for almost in the same

Engraving of Giambattista Porta in his laboratory, from his book, Magia Naturalis.

breath he proclaimed that he could himself build a burning mirror that would set objects ablaze at an infinite distance:

> I will show you a far more excellent way than the rest . . . it exceeds the invention of all the ancients, and of our age also, and I think the wit of man cannot go beyond it.

And what did Porta plan to do with such an all-powerful mirror? "Cast forth terrible fires and flames that are most profitable in warlike expeditions," he fantasized, fires that would burn "the enemy's ships, gates, bridges and the like." As far as we know, Porta never built this stupendous weapon—perhaps it was as mythical as his recipe for producing silver by combining quicksilver with a toad in a simmering pot!

During the sixteenth and seventeenth centuries, many scholars and scientists speculated about the power of large mirrors. Here, a series of mirrors focuses sunlight to destroy a distant building.

The grandiose claims of Porta and Cardano led Athanasius Kircher, a Jesuit scientist of the seventeenth century, to "press with all zeal" his hunt for such a giant mirror. While Kircher could not deny that if such "a parabolic mirror were the size of a mountain, it would burn objects at great distances," he wondered who could manufacture such a prodigious machine. To obtain the answer, Kircher traveled throughout Europe, meeting with outstanding craftsmen, "so that they might display anything similar" to the huge burning mirrors he had heard about. But after an extensive search, Kircher came away empty-handed. "Nowhere has such [a powerful mirror] as I sought appeared," he wrote.

Kircher also tested and had others test the smaller mirrors that he did find. All of them failed to meet the claims of their inventors. For instance, a mirror built by Manfredo Septala of Milan was reputed to quickly reduce to burning embers a piece of wood placed 15 to 16 paces (about 40 feet) away. On Kircher's instructions, an independent investigator tested this claim. He wrote back that yes, it was possible to ignite the piece of wood at that distance, but it required the time it took to say a long Miserere—a very involved Catholic prayer!

Kircher ended his report with stern and sober words: "Let not mathematicians boast about more things than they can demonstrate, and let them not expose themselves and the noblest art to mocking and jests of men." Kircher's advice was apparently heeded—or perhaps the Jesuit merely echoed the growing empirical sentiment of the seventeenth century. In any event, the claims of mirror builders in later years were more in line with results that could be verified by experiment.

The development of powerful burning mirrors proceeded slowly. In the late 1600's, an optician from Lyons, Villette, built several large spherical mirrors. The largest measured more than three feet in diameter—about a third larger than any mirror previously built. As to the solar heat generated by one of these mirrors, an observer commented, "One may pass one's hand through it [the focal point], if it be done nimbly; for if it stay there the time of a second, . . . there is danger of receiving much hurt.'' Another writer declared in an article published in London that a Villette mirror was able to make tin melt in three seconds, cast iron melt in 16, and a diamond lose 87 percent of its weight. ''In short,'' he concluded, ''there is hardly any body which is not destroyed by this fire.''

Mirror Technology Improves

Although Villette's mirrors were larger than those of his predecessors, they were nowhere near as large as those proposed by Bacon or da Vinci. Reflector size was still limited by the use of heavy, brittle alloys. But late in the seventeenth century, other methods and materials were developed to overcome these problems, and mirrors were produced that were both lighter and easier to handle.

A Dresden mechanic named Gartner constructed mirrors from wood coated on its concave inner surface with wax and pitch and then a layer of shiny gold leaf. But wooden mirrors had their own shortcomings. ''Not only do these mirrors not bear the circumstances that go with burning such as flying sparks, flowing streams of molten metal and slag,'' one expert pointed out, but the gold leaf easily deteriorated—completely destroying the mirror's reflectivity.

A German noble, the Baron of Tchirnhausen, successfully tried another material more durable than wood and lighter than alloy metal—copper. He hammered a sheet of copper ''scarce twice as thick as the back of an ordinary knife'' into a mirror the like of which ''hath not yet been made . . . for in magnitude it exceeds even the great one [Villette's mirror] which they shew as a sight in Paris.'' This was no exaggeration. The Baron's device was the largest burning mirror of the seventeenth century—five and a half feet in diameter. Yet this mirror was still quite manageable because it was made from a fairly thin, relatively lightweight sheet instead of the heavier cast alloy. Moving it to track the sun could be easily handled by one man.

This solar reflector was the most powerful burning mirror of the century. The Baron wrote:

> The force of this speculum [mirror] in burning is such [that] even chemists who best know the power of fire will hardly credit their own eyes. . . . A piece of tin or lead three inches thick, as soon as it is put into the focus, melts away in drops. . . . A plate of iron or steel placed in the focus immediately is seen to be red hot on the back side; and soon after a hole is burnt through.

But the Baron's brilliant invention had its shortcomings; it was still difficult to fabricate a single sheet of copper large enough to make a mirror of such dimensions.

A different tack was taken by Peter Hoesen, an eighteenth-century mechanic and Dresden's royal carpenter. So far all methods of mirror building had been inherently restrictive because they basically used a single piece of material—whether alloy, wood or copper plate. Hoesen abandoned unitary construction and built his mirrors from sections of hard wood covered with pieces of brass. Hoesen carved a parabolic shape from a skeleton made of cross members of very durable wood, and lined the inner concave surface with strips of brass sheet metal measuring 5 ft 5 in. by 2 ft 8 in. ''With skillful use of his hands,'' he explained, ''he fits it [the brass] to the contours of the depression so that the seams between each sheet are hardly to be perceived.''

Large burning mirror of the late 1700's. Like Peter Hoesen's model, this mirror was built in sections.

This method enabled the German craftsman to build mirrors as large as ten feet in diameter—almost three times the size of Villette's big reflector, and almost twice as large as the Baron's. Never before had Europeans witnessed such a colossal burning device. These mirrors, by far the most powerful solar reflectors yet developed, concentrated the rays of the sun in a target area less than one inch across. "The hardest stones," said Hoesen, "resist [their] force for only a few seconds. Things of a vegetable nature burst into flames immediately, turning in a short time to ash. . . . The bones of animals become calcined." The power of these mirrors was verified by a research worker who experimented with a Hoesen reflector five feet in diameter. He discovered that copper ore melted in one second, lead melted in the blink of an eye, asbestos changed to a yellowish-green glass after only three seconds, and slate became a black, glassy material in 12 seconds.

Despite their size, Hoesen's mirrors were easy to handle. "One can put it in all positions using but one hand!" remarked an observer. The mirror's mobility was due to its light weight and also to the clever design of its mounting system. The reflector was supported by two semicircular wooden arms with adjustable screws so that the device could be tilted in many different directions. The arms met at the base of a tripod mounted on rollers, making the entire mechanism mobile.

Although Hoesen built many burning mirrors of substantial size, Bacon's ominous prophecy was never realized—none were employed as weapons of war. By now gunpowder had given humankind a far more efficient and versatile means of delivering death and destruction over long distances. For a long time burning mirrors remained in the hands of scientists who still experimented and tinkered with them without putting them to any practical use. Not until the following century—during the Industrial Revolution—was da Vinci's dream of using burning mirrors to produce solar power for industry revived.

A conservatory in Victorian England. Early in the nineteenth century, these glassed-in plant showcases became popular additions to many upper-class homes.

Chapter 4
Heat for Horticulture

Just as the ancient technology of burning mirrors was lost to Europeans during the Middle Ages, the use of glass windows to trap solar heat was almost unknown in the West for many centuries after the Fall of Rome. Window glass was not as common as during the heyday of the Roman Empire. People lived in the midst of constant turmoil—caught between local lawlessness and wars among rival feudal lords and kings. In most regions no central authority existed that could effectively keep the peace and afford protection to its citizens. Hence, homes had to be built as defensible units. Easily breakable glass windows gave way to small chinks in the walls of houses to let in light and air. Usually the church was the only building in a medieval village to have glass windows, but they were made of stained glass.

Economic and social conditions discouraged the use of glass in greenhouses as well. Times were hard. Only a small minority could afford such luxuries as raising exotic plants in greenhouses, and specimens from abroad were hard to obtain because trade with distant lands had slackened off. Moreover, the threat of periodic warfare did not permit even the rich to devote much attention to such frivolities. Greenhouses also met with strong opposition from the Church. Just as Bacon's idea of turning a gigantic solar mirror into a doomsday weapon was suppressed because it was seen as demonic tampering with the divine plan, so the Church denounced the growing of plants apart from their natural habitat or out of season. According to legend, a twelfth-century Dominican friar who experimented with forcing fruits and flowers in a greenhouse as the Romans had done was burned at the stake for practicing witchcraft.

Horticulture's Revival

Collecting solar heat for horticulture enjoyed a revival during the sixteenth century. The tide of empirical science had begun to break the bonds of Church dogma. The growing wealth and stability of Europe contributed to the favorable atmosphere. With exploration and trade came an increased flow of money and an appetite for more comfortable living. Ships were now returning from Asia, Africa, and the New World with beautiful flowers such as African violets and delicious foods such as bananas, pineapples, and coffee. People wanted to grow these exotics at home and enjoy native fruits and vegetables in all seasons—just as the Roman Emperor Tiberius had satisfied his year-round craving for cucumbers by raising them in glazed cold frames more than a thousand years before.

The Dutch and Flemish were the first modern northern Europeans to develop horticulture to a level equaling or surpassing that of the Romans. Perhaps their early independence from the authority of the Church encouraged their pioneering efforts in the field of scientific gardening. Certainly their great success in world trade helped to provide the new bourgeois merchant class with the means to take up such gentlemanly pursuits as raising exotic plants. J.C. Loudon, a late eighteenth century horticulturalist, observed that "Horticulture . . . was in great repute in all the low countries during the seventeenth century."

The French and English followed suit, trying to grow plants from the warmer regions of the world in the inhospitable climate of northern Europe. They also sought to improve the yield of native crops and to grow them out of season. Experts wrote books for the amateur gardener on how to grow "all the delicacies that belong

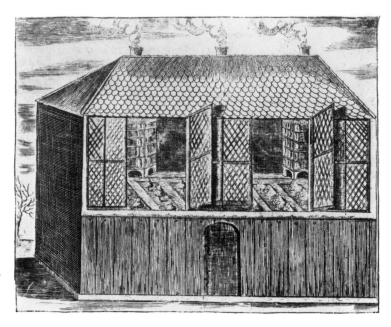

Early Dutch greenhouse, circa 1550. Ceramic woodburning stoves at rear helped to keep it warm in winter. (Reprinted by permission of the Houghton Library, Harvard University.)

to every country and to accelerate their ripening, so as to obtain throughout the whole of the year the use of those fruit which nature affords . . . only in summer.''

The unusually short growing seasons during this era were a major problem for gardeners. The years 1550 to about 1850 have been called ''The Little Ice Age'' because Europe suffered extremely short summers and very severe winters. In England, for example, the average temperature from 1680 to 1719 ranged from 58.6°F in summer to 37.9°F in winter. The cold was so extreme that on 13 occasions during the seventeenth and eighteenth centuries the ice on the Thames could support the weight of people—a rare occurrence during the previous five centuries.

Only if ''nature was assisted by art,'' as one gardening book put it, could a lover of plants ''cure this great evil and dangerous enemy,'' the cold. Solar energy became a favorite tool in this battle. Its power was extolled by Joseph Carpenter in 1717:

> The sun by its heat dissipates the cold and gross humours of the earth; it renders it more refin'd and easier for the vegetation of seed and fruit trees. 'Tis by the influence of this noble planet that the sap rises up between the wood and bark, producing the first buds, then the leaves and fruits; its beams serve not only to ripen the fruit, but it makes it large, beautiful to the eye, and pleasant to the taste.

The art of harnessing solar heat, of learning how to enhance its beneficent effects, was evolving once again.

Fruit Walls

A technique devised by the French and English to make better use of solar energy in their gardens was to grow plants near a fence or wall heated by the sun. This fence or wall would later release its heat to the plants. The branches of fruit

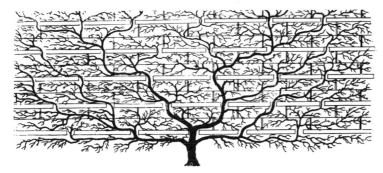

Branches of a fruit tree were attached to a brick "fruit wall" that collected the sun's heat and hastened the ripening process.

trees were even nailed to sun-heated walls to help ripen the fruit more quickly. One gardening tract pointed out that this practice resulted in substantially greater production and avoided "our entire want [in] some years of the best and latest fruit" of the season.

As for the heat-absorbing qualities of various kinds of fruit walls, most experts recommended brick. Nicolas Fatio de Duillier, author of the influential 1699 essay *Fruit Walls Improved*, explained why:

> As to the properest matter of our walls, I think brick to be much better in this countrey [England] than stone: because they grow much hotter, and keep much longer the heat. By which means they do still warm the plants a good while after the sun is hid under a cloud and in a manner lost to other walls.

Early fruit walls were usually built perpendicular to the ground facing south. But these types of walls had their shortcomings. A vertical wall received the sun's rays on its south face for only a few hours in summer. Because the sun was then high overhead, its rays shone obliquely on the wall, not directly. Fatio de Duillier discussed the difficulty:

> When the days are something long, and the heat of the summer is in its greatest strength, it is late before the sun shines upon them [fruit walls], and the sun leaves them as early in the afternoon. When it is about midday the sun is so high that it shines but faintly and very sloping upon them, which makes the heat much less, both because a small quantity of rays falls then upon these walls; and because that very quantity acts with a kind of glancing, and not with full force.

One solution was to construct fruit walls toward the southeast, rather than directly south. As noted by Stephen Switzer, an eighteenth-century gardening authority, "the sun shines early on [the southeast wall] . . . and never departs from it till about two o'clock in the afternoon." Even so, a vertical southeast wall lacked sunshine during the remainder of the day. Semicircular walls called "half-rounds" were not much better. Although "every part of the wall, one time of the day or the other, [enjoys] a share of the sun; and the best walls will not fail of being exceedingly hot by the . . . collection of the sun beams," according to Switzer, each section of the wall still had access to the sun for only part of the day.

Perhaps the most ingenious solution was the "sloping wall" described by Fatio de Duillier. A south-facing perpendicular wall is exposed to only one half of the visible sky, but a wall built at an incline of 45° from the northern horizon and facing

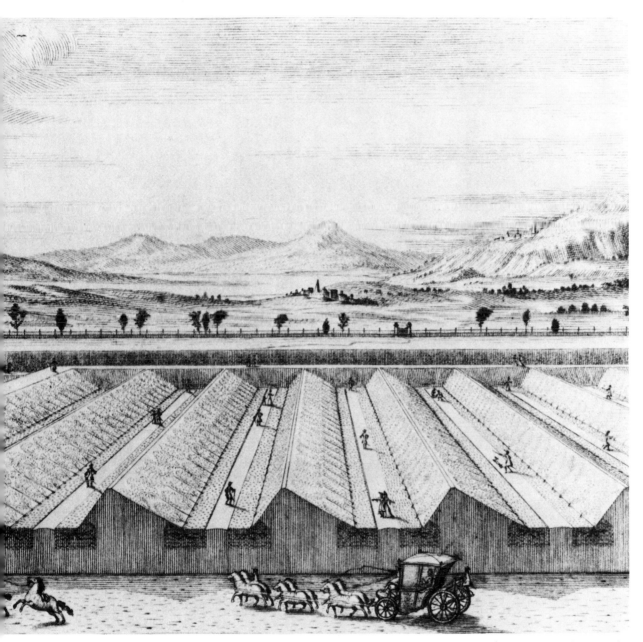

Engraving of sloping fruit walls in seventeenth-century France, from Fruit Walls Improved, *by Nicolas Fatio de Duillier. These walls faced the southern horizon at an angle chosen for optimum year-round solar collection.*

south is exposed to three quarters of the sky and can absorb the sun's energy for a longer part of the day. A sloping wall also receives more direct solar rays and therefore gets hotter. Fatio de Duillier calculated that in England at the summer solstice, the intensity of solar energy striking a sloping wall is 3½ times greater than that hitting a perpendicular wall facing the same direction.

Fatio de Duillier confidently predicted that the increased heat gain of sloping walls would help produce "grapes and figs and other fruits equal here [England] in goodness to those of some much hotter climates." Indeed, the French had cultivated fine crops of grapes for years by growing them on sloping walls facing south. Fatio de Duiller even suggested building a pivoting fruit wall that would follow the sun. Such a wall would "be sure to enjoy almost all the sun's heat."

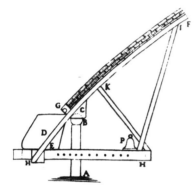

Above: Fatio de Duillier's proposal for a "tracking" fruit wall, 1699. This mechanism pivoted about an axis (A-B) and followed the sun's daily motion across the sky.

Right: English horticulturalists commonly used glass "cold frames" to extend the growing season. Solar heat trapped by the glass covers allowed for the enjoyment of exotic plants grown out of season.

Greenhouses Come of Age

In 1714 the Duke of Rutland tried to use sloping walls in his English garden during the winter, but found that they did not provide enough solar heat to keep his plants alive. He therefore placed glass frames over the walls to keep the collected solar heat from dissipating so quickly. The use of glass as a solar heat trap was not entirely new to England. Some forty years earlier, Sir Hugh Plat suggested that his patrons use glass to protect new seedlings because it would "defend off the cold air and increase the heat of the sun." Cold frames and glass greenhouses soon became immensely popular in England, as well as in Holland—where Europe's first modern greenhouses had been built in the 1500's—and across the continent. In fact, many have called the eighteenth century the "Age of the Greenhouse" because it became fashionable for nearly every man of means to have one.

New glass-manufacturing methods allowed the production of large windows for greenhouses and homes. Previously, from the eleventh century to the end of the seventeenth century, window panes were usually made by the crown glass method. A craftsman blew the hot glass into a bubble, and thrust a rod into the top of the bubble directly opposite the blowpipe. He detached the blowpipe, leaving an air hole where the tool had been, and as the glass began to cool he reheated the bubble and twirled the rod until centrifugal force caused the bubble to flatten into a disc. When the disc had cooled and hardened, the glass was cut into small, thin panes.

Locating crown-glass factories near fuel-rich areas helped to lower the price of glass—making windows more readily available to members of the ascendant middle class. Their growing demand for larger and thicker panes was finally met when the French developed the plate glass process at the end of the seventeenth century. This process was strikingly similar to the Roman method. Glass was melted in a large cauldron, and several workmen carried the molten liquid to a casting table where they poured it into a rectangular frame. With an iron roller, they flattened the glass to a standard thickness. After the plate had cooled and hardened, it was ground and polished. This plate glass method produced window panes measuring up to six feet on a side, and it quickly eclipsed the old crown glass process— although crown glass remained common in England until the end of the eighteenth century.

Innovative Greenhouse Designs

Scientists sought novel greenhouse designs to enhance their solar heat collection and storage abilities, because they wanted to reduce the amount of fuel needed to keep plants warm at night, on cloudy days and in the winter. They hoped to save fuel and believed that plants in a solar-heated greenhouse grew better than plants raised in artificial heat.

As with fruit walls, early greenhouses were built with a southern exposure. Scientists soon realized that greenhouse walls should also be sloped to capture more sunlight. Hermann Boerhaave, a seventeenth-century Dutch professor of botany, demonstrated that in Holland's northern latitude the rays of the low-lying winter sun would enter a greenhouse more directly if the glass walls were quite steeply inclined. He determined that an angle of 75.5° from the northern horizon

Top: Eighteenth-century engraving of the crown glass method, from the French Encyclopedia of Denis Diderot and Jean d'Alembert.

Above: The plate glass process, developed by French glassmakers in the late 1600's. Much larger window panes could be fabricated with this method.

would be best for a latitude of 52.5°. Most directors of botanic gardens in Europe took Boerhaave's advice and built their greenhouses accordingly. Greenhouses intended primarily to encourage the growth of larger and more flavorsome fruit during the summer required a less acute angle of incline, because the sun's course was more directly overhead. Using this strategy, one English gardener reported "the most abundant crops of grapes perfectly ripened with less time and effort and less expenditure on fuel than I have witnessed in any other instance."

Michael Adanson, an eighteenth-century French scientist, recommended that

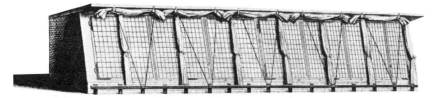

Top: Canvas curtains were used for nighttime insulation on this eighteenth-century Dutch greenhouse. During this period, the Dutch also used double pane glass to help control heat losses.

Above: Interior view of a seventeenth-century Dutch greenhouse.

the floor of the greenhouse rather than its walls be sloped. His approach was much akin to placing a glass cold frame over a sloping fruit wall. Adanson wrote the first systematic treatise on the theory and construction of greenhouses. He presented rules, tables, and diagrams to be followed for building the most functional greenhouses in every possible location, from the poles to the equator.

Gardeners not only sloped the walls or floors of their greenhouses, but they also invented insulating techniques to increase heat retention. When the sun was not shining, they placed mats or canvas coverings over the greenhouse to conserve the solar heat that had been collected—and also the heat generated by fires that were often lit inside. The Dutch constructed greenhouses with two layers of glass—the dead air space between the layers acting as insulation.

Late in the eighteenth century, Dr. James Anderson took the idea of solar heat storage a step further. Normally when the sun was shining and the greenhouse became too hot, some of the solar-heated air was released by opening windows in the structure. In Anderson's novel design, this hot air was captured and stored for later use. He divided his greenhouse into an upper and lower chamber. During the day hot air collecting in the lower chamber rose through a pipe to the upper. At

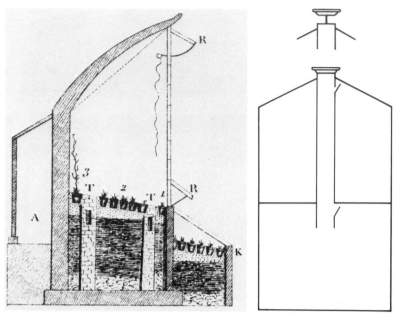

Left: Michael Adanson suggested sloping the planting beds, rather than the greenhouse glass, to increase exposure to the winter sun.

Right: Patent drawing of a greenhouse with solar heat storage, developed by James Anderson in 1803. Solar-heated air was stored in the upper chamber by day and returned to the lower chamber at night.

night, cold outside air was admitted to the upper chamber, forcing the stored hot air through a duct back into the lower chamber—to heat the plants. How well this storage system worked cannot be ascertained from the records, but it was an early attempt to store solar heat long enough so that it could be used when the sun was not shining.

The Development of Conservatories

As wealth accumulated during the nineteenth century in England and in other European countries, the greenhouse began to assume a more lavish form—the conservatory. In this glassed-in garden, the well-to-do could leisurely amble with their guests through lush, jungle-like foliage. *The Gentleman's House*, an architectural guide for the wealthy landowner, pointed out the difference between the greenhouse and the conservatory: "The greenhouse is a structure in which plants are cultivated as distinguished from the conservatory in which they are placed for display." The greenhouse was primarily functional whereas the conservatory was a place where exotic plants were put on display for atmosphere. Gardening manuals described the fuel-saving features of a southern orientation for the conservatory.

Architects such as the British designer Humphrey Repton brought the sunlit ambience of the conservatory right into the home by attaching this glass garden onto the south side of a living room or library. On sunny winter days the doors separating

Early solar remodeling. Architect Humphrey Repton argued that a dull interior (top view) could be transformed into a vibrant home by adding an attached conservatory (lower view).

the conservatory and the house were opened to allow moist, sun-warmed air to circulate freely into the otherwise gloomy, chilly rooms. Some contended that the conservatory also gave families a healthier way of spending their free time. As *The English Gardener* exclaimed, "How much better during the long and dreary winter for daughters and even sons to assist their mother in a greenhouse than to be seated at cards or in the blubberings over a stupid novel!''

By the late 1800's the country gentry had become so enamored of attached conservatories that they became an important architectural feature of rural estates. According to John Hix, author of *The Glass House*, the conservatory was ''no longer . . . seen as a simple extension to the dwelling, but an integral way of life.'' As one gardening manual of the period noted, ''It is most usual to connect a conservatory with either the drawing room; or what is probably better than all, with a saloon, vestibule, gallery, or corridor immediately adjoining one or more apartments.'' Conservatories also found enthusiasts in the northeastern United States, where the wealthy could enjoy a stroll through their verdant wonderlands even in the harshest months.

This fashion filtered down to members of the middle class, but on a much smaller scale. Modest conservatories tightly hugging the outer shell of the home were common in urban areas. Rooftops of multistoried buildings also served as sites for greenhouses. As one writer remarked, ''a warm greenhouse on the roof [is] a more pleasant thing than a dark parlor.'' For the crowded city flat, a large window garden on the south wall had to suffice. Jacob Forst, a leading British horticulturalist, envisioned that the south side of every urban building could be glassed over for growing grapes, figs, and cherries. ''Such walls would never need paint,'' Forst argued, and would offer an ''admirable arrangement for house ventilation'' by trapping sun-heated air and circulating it to the interior of the building.

As conservatories became popular, people grew indifferent to the direction in which they faced. Instead of the sun's rays streaming in from the south, artificial heating systems now provided warmth for the garden houses, and conservatories became fuel consumers rather than fuel savers. One of the chief reasons for the

Noble pleasures in Victorian England. Solar-heated air from the conservatory often warmed the adjoining rooms of a house.

demise of the conservatory in England was the institution of fuel rationing during World War I. The lesson of solar heating had been discovered and then lost.

Large glass windows on the south side helped plants flourish in this London setting. A rooftop greenhouse provided further growing space.

Horace de Saussure, a European scientist who developed the hot box in 1767.

Chapter 5
Solar
Hot Boxes

The increased use of glass during the eighteenth century made many people aware of its ability to trap solar heat. As Horace de Saussure, one of Europe's foremost naturalists of the period, observed: "It is a known fact, and a fact that has probably been known for a long time, that a room, a carriage, or any other place is hotter when the rays of the sun pass through glass." This French-Swiss scientist was quite surprised that such a common phenomenon had not led to any empirical research on the maximum temperature attainable in a glass solar heat trap. When experimenting with solar energy, his contemporaries preferred to work with burning mirrors, which could perform such amazing feats as burning objects at a distance or melting the hardest metals within seconds. In 1767, de Saussure set out to determine how effectively glass heat traps could collect the energy of the sun.

De Saussure first built a miniature greenhouse five walls thick. He constructed it from five square boxes of glass, decreasing in size from 12 in. on a side by 6 in. high to 4 in. on a side by 2 in. high. The bases of the boxes were cut out so the five boxes could be stacked one inside the other atop a black wooden table. After exposing the apparatus to the sun for several hours, and rotating the model so that solar rays always struck the glass covers of the boxes perpendicularly, de Saussure measured the temperature inside. The outermost box was the coolest, and the temperature increased in each succeedingly smaller box. The bottom of the innermost box registered the highest temperature—189.5°F. "Fruits . . . exposed to this heat were cooked and became juicy," he wrote.

De Saussure seemed unsure of how the sun heated the glass boxes:

> Physicists are not unanimous as to the nature of sunlight. Some regard it as the same element as fire, but in the state of its greatest purity. Others envisage it as an entity with a nature completely different from fire, and which, incapable of itself heating, has only the power to give an igneous fluid the movement which produces heat.

Despite de Saussure's shaky theoretical underpinnings, the validity of his test results is beyond question.

Today we can better explain what went on in de Saussure's glass boxes or what will occur in any glass container or glass-walled building exposed to the sun. Sunshine penetrated the glass covers of the boxes, and was absorbed by the black surface of the table on which the boxes rested. In the process, the light energy was converted into heat. Much of this heat was released into the glass boxes as warm air and thermal radiation. But clear glass has a peculiar property: it easily allows sunshine to pass through, but inhibits thermal radiation from doing the same. Therefore this trapped energy heated the air inside the box. The glass walls also blocked the heated air from escaping, but some heat was lost by conduction through the glass.

Building a Better Heat Trap

Seeking to block the heat loss even more effectively, de Saussure made a small rectangular box out of half-inch pine and lined it with black cork. Three separate sheets of glass covered the top of the box. When exposed to the sun, the bottom of the box reached a temperature of 228°F—16°F above the boiling point of water, and

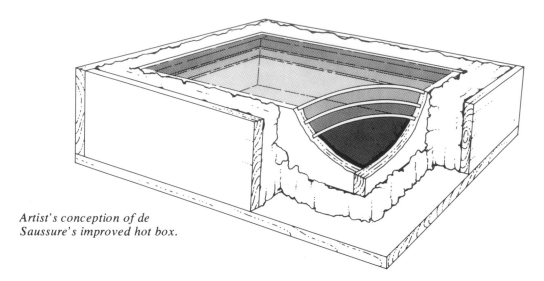

*Artist's conception of de
Saussure's improved hot box.*

almost 40°F higher than in the first experiment. This device was later called a *hot
box* because of the large amount of solar heat it could retain.

However, the hot box was still losing some heat to the outside. De Saussure
therefore placed the wooden box into the middle of an open-topped container and
stuffed wool packing between the sides of the container and the walls of the box.
The added insulation kept more heat inside, and the temperature in the hot box
reached 230°F even though the weather was not as favorable as during the prior
experiment.

The hot box helped de Saussure ascertain why it is cooler in the mountains than
in lower-lying regions. His hypothesis was that the same amount of sunlight strikes
the mountains as the flat lands, but because the air in the mountains is more
transparent it cannot trap as much solar heat. To test the theory, de Saussure carried
a hot box to the top of Mt. Cramont in the Swiss Alps. The thermometer in the hot
box hit 190°F, while the temperature outside was 43°F. The following day he
descended to the Plains of Cournier, 4,852 feet below, and repeated the experiment.
Although the air temperature was 34°F hotter than on the mountain, the temperature inside the hot box was almost the same as in the previous experiment.

Thus de Saussure's hypothesis was confirmed: the sun shines with almost equal
force at higher and lower elevations—as proved by the equal temperatures in the
hot box on the mountain and on the plains. It was the difference between the
atmosphere in the mountains and on the plains that caused the difference in outdoor
air temperature. At lower elevations there are greater amounts of carbon dioxide
and water vapor in the air. This denser atmosphere holds in the solar heat more
effectively, retarding its escape into space; so it gets hotter at these elevations. But
the glass covers of a hot box present an equally effective barrier to solar heat trying
to escape from the box whether it is located in the mountains or at sea level; so it
registers the same temperature in either place.

De Saussure's hot box served as a model for nineteenth-century scientists
demonstrating the relationship of the sun to the earth and its atmosphere. Like the

Portrait of Sir John Herschel. On an expedition to South Africa in the 1830's, this eminent British astronomer built a hot box and used it to cook meals.

glass covers of the hot box, our atmosphere allows most sunlight to strike the earth. About three quarters of the sun's radiation reaches the earth's surface when the sky is clear. The earth, like the bottom of the hot box, absorbs sunlight and releases heat. But this heat cannot readily escape through the atmospheric blanket—just as solar heat is trapped by the panes of glass in a hot box.

Later Hot Box Experiments

Several nineteenth-century scientists conducted experiments with hot boxes and obtained comparable results. Sir John Herschel, the noted astronomer, made a hot box while on an expedition in the 1830's to the Cape of Good Hope in South Africa. It was a small mahogany container blackened on the inside and covered with glass, set into a wooden frame protected by another sheet of glass and by sand that was

Samuel Pierpont Langley experimented with a hot box on an 1881 climb of Mount Whitney, California. He later became Director of the Smithsonian Institute.

heaped up along its sides. The outcome of Herschel's experiments with this hot box was not only scientifically interesting but also pleasing to the palate, as his notes indicate:

> As these temperatures [up to 240°F] far surpass that of boiling water, some amusing experiments were made by exposing eggs, meat, etc. [to the heat inside the box], all of which, after a moderate length of exposure, were found perfectly cooked. . . . [On] one occasion a very respectable stew of meat was prepared and eaten with no small relish by the entertained bystanders.

The following day, using a much simpler hot box, Herschel made an egg,

> Which burned Peter's [his son's] fingers as if fresh from the pot. It was done as hard as a salad egg and I ate it and gave some to my wife and six small children that they might have it to say they had eaten an egg boiled hard in the sun in South Africa.

The story of Sir John's solar cookouts intrigued Samuel Pierpont Langley, the American astrophysicist who later became head of the Smithsonian Institution. Langley had been fascinated by solar heat ever since he was a child, when he wondered why glass kept the interior of a greenhouse warm. In 1881, Langley took a

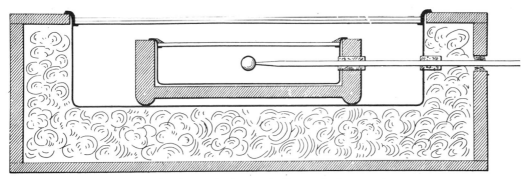

Cross-section of Langley's hot box, which was similar to de Saussure's later models. A thermometer penetrating the walls at right was used to measure the air temperature inside the inner box.

trip to Mt. Whitney to study the effects of solar energy. There he experimented with a hot box. He related his experiences in an 1882 issue of *Nature*:

> As we slowly ascended . . . and the surface temperature of the soil fell to the freezing point, the temperature in a copper vessel, over which lay two sheets of plain window glass, rose above the boiling point of water, and it was certain that we could boil water by the solar rays in such a vessel among the snow fields.

De Saussure, Herschel, and Langley all demonstrated that temperatures exceeding the boiling point of water could be produced in a glass-covered box. Its inventor realized that the hot box might have important practical applications. As de Saussure stated almost self-effacingly, "Someday some usefulness might be drawn from this device . . . [for it] is actually quite small, inexpensive, [and] easy to make." Indeed, his modest hope was more than fulfilled: the hot box became the prototype for the solar collectors of the late nineteenth and twentieth centuries—collectors that were able to supply hot water and heat for homes and provide power for machines.

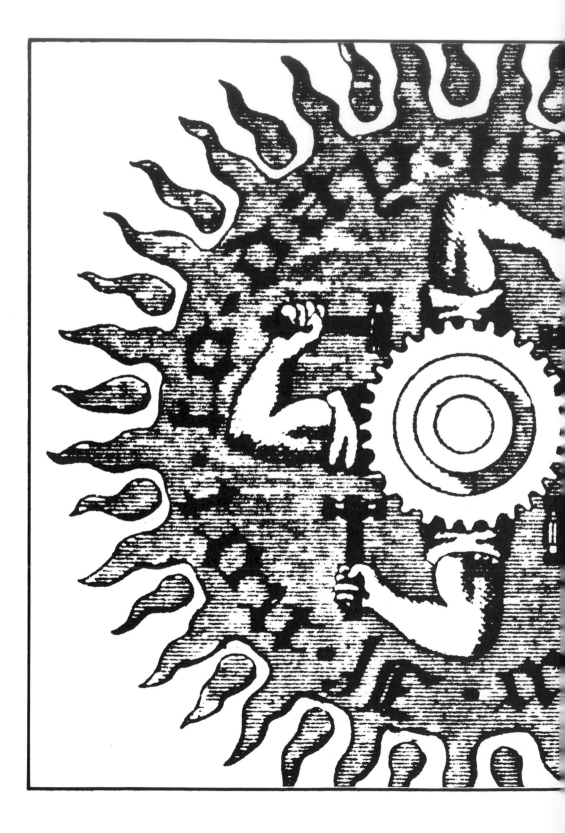

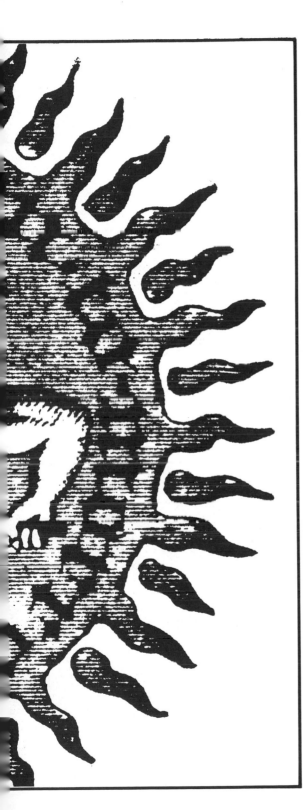

II

Power from the Sun

Augustin Mouchot's largest sun machine, on display at the Universal Exposition in Paris, 1878.

Chapter 6
The First Solar Motors

By the early 1800's the sporadic advances of science and technology in previous centuries had begun to snowball, leading to a full-scale Industrial Revolution. The use of machines to augment the muscle-power of men and animals meant that goods could be manufactured on an unprecedented scale. But mechanization depended on the production of iron, and to make one ton of iron took seven to ten tons of coal. Coal, in addition to wood, was also in demand as a major fuel source to power the newly developed steam engines and supply heat for the factories springing up throughout Europe.

France was at a disadvantage compared to other industrial countries because she had to import almost all of her coal. As a consequence, she lagged far behind rapidly industrializing England. So the French decided to pursue an aggressive program to step up domestic coal production. The plan worked, and output doubled in two decades—providing resources and power for iron smelters, textile plants, flour mills, and the many other new industries that began to appear by the second half of the nineteenth century.

Many Frenchmen now felt secure about the nation's energy situation. But not everyone shared this complacency. In 1860 Augustin Mouchot, a professor of mathematics at the Lyceé de Tours, cautioned:

> One cannot help coming to the conclusion that it would be prudent and wise not to fall asleep regarding this quasi-security. Eventually industry will no longer find in Europe the resources to satisfy its prodigious expansion. . . . Coal will undoubtedly be used up. What will industry do then?

Mouchot's answer was, "Reap the rays of the sun!" To show that solar power could be harnessed to run the machines of the Industrial Age, he embarked upon two decades of pioneering research.

Early Uses of Sun Power

Mouchot began investigating the potential of solar machinery with a study of its historical roots. His findings surprised his contemporaries, who thought that solar power was a new concept. As Mouchot put it,

> One must not believe, despite the silence of modern writings, that the idea of using solar heat for mechanical operations is recent. On the contrary, one must recognize that this idea is very ancient and in its slow development across the centuries it has given birth to various curious devices.

The first of these "curious devices" powered by the sun was built by Hero of Alexandria in the first century of the Christian era. Hero invented a solar syphon that could transfer water from one container to another when it was placed in the sun. Solar energy heated air inside a closed sphere; the heated air expanded and exerted pressure on water inside the sphere, forcing it out.

During the sixteenth and seventeenth centuries, many natural philosophers proposed solar machines based on the same principle. Athanasius Kircher, the Jesuit priest who searched the continent looking for giant solar reflectors, claimed he had developed a solar clock—although it is not clear whether his complex

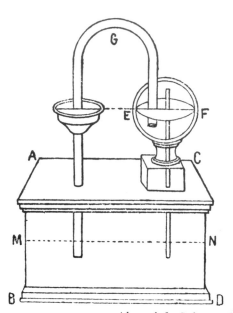

Above left: Solar syphon built by Hero of Alexandria, first century A.D. Solar-heated air in the globe (E-F) expanded, forcing water out a tube (G).

Above right: Solar whistle developed by de Caus, 1659. Working on the same principle as Hero's syphon, this device sounded when the sun shone upon it.

inventions really worked. In his book *New and Rare Inventions of Water Works*, published in 1659, Isaac de Caus told how to make:

> An admirable engine, the which being placed at the foot of a statue, shall send forth sound when the sun shineth upon it, so as it shall seem that the statue makes the said sound.

This device imitated the legendary "voice of Memnon," a wailing sound which issued from a Theban statue of the Ethiopian king Memnon when the morning sun struck it. The ancient historian Tacitus and the Greek geographer Pausanias described this wonder in their writings. De Caus recreated this voice by connecting a solar syphon to a whistle. The syphon consisted of two adjacent metal boxes made of copper or lead. One box was partly filled with water and the other was empty. As the sun's rays heated the water-filled box, the air inside expanded and forced the water out of the container through a curved tube and into the second box. The water gradually displaced the air in the second box, and as the air flowed upwards into two organ pipes atop the apparatus, it produced sound.

Mouchot's First Attempts

Although such solar inventions aroused Mouchot's curiosity, he complained that "no one has adapted them in a practical way." A true child of the Industrial Age, he was not content to see solar energy simply used for amusing contraptions. The

This solar pump designed by Isaac de Caus used a series of lenses to focus sunlight onto tanks of water.

practical development of solar power to serve industry became his principal pursuit.

Mouchot was 35 when he began his research at Tours in 1860. His objective was to find a way to collect the sun's energy efficiently enough to drive industrial steam engines economically. A hot box resembling de Saussure's seemed promising because it could generate temperatures high enough to produce steam. But Mouchot's initial experiments proved disappointing. He felt that a hot box large enough to run an industrial machine would take up a great deal of room and be much too expensive.

Mouchot's second solar collector was also based on the hot box concept, but its design provided greater exposure to the sun's rays. A bell-shaped copper cauldron coated on the outside with lamp black was covered by concentric glass bell jars "to retain, as in a trap, the heat of the sun." Because a 360° glass surface replaced the glass top and wood walls of the old-style hot box, the solar collection area was greater. At all times of the day, the sun struck some part of the bell jar perpendicularly. By contrast, the hot box had to be moved constantly to keep its glass top

oriented toward the sun. Mouchot found that the apparatus could collect "practically all the rays falling upon the exterior bell, which is to say a rather large sum of heat relative to the volume of the apparatus." Nevertheless, an impractically large device was still required to produce enough heat for industrial purposes.

The solution Mouchot tried next ingeniously combined two solar developments that had thus far evolved independently: the glass heat trap and the burning mirror. A solar reflector could concentrate more sunlight on the collector than the collector could receive on its own. The glass heat trap could then be kept to a manageable size and still produce sufficient heat to drive engines. Thus, Mouchot considered a mirror "indispensable to making the solar device practical." Linking the two approaches led to several successful inventions: a solar oven, a solar still, and a solar pump.

The solar oven had a tall, blackened cylinder of copper surrounded by a cylinder of glass, with a one-inch air space in between. Food went inside the copper cylinder, which was then covered by a wooden lid. The solar mirror was shaped like a vertical trough; it faced south and reflected a band of sunlight onto the north side of the cylinders. The mirror was made of polished silver sheets attached to a wooden frame. Mouchot cooked "excellent" dinners in this apparatus, just as Herschel had done several decades earlier. Mouchot claimed that:

> This new oven allowed me, for example, to make a fine pot roast in the sun. This pot roast was made out of a kilogram of beef and an assortment of vegetables. At the end of four hours the whole dinner was perfectly cooked, despite the passage of a few clouds over the sun, and the stew was all the better since the heat had been very steady.

With a few modifications Mouchot converted the solar oven into a still that could make wine into brandy. Whereas stills of the period normally relied on coal or

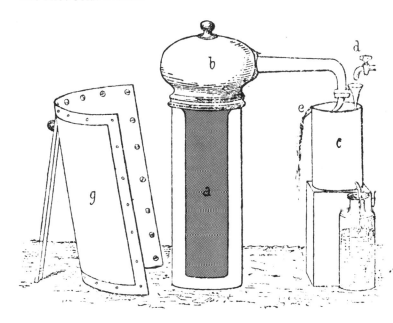

Opposite page, left: Mouchot's first solar cooker, as illustrated in his book La Chaleur Solaire. *A blackened copper cylinder (b) covered by a glass sleeve absorbed sunlight reflected onto it by a mirror (d).*

Opposite page, right: Cross-section of Mouchot's first solar pump, patented in 1861. Sunlight heated air inside a copper cauldron (B); this air expanded and drove water (A) out of the base through an escape valve (C).

Left: Solar still built by Augustin Mouchot. Alcohol evaporated in the copper cauldron (a), and its vapor was delivered to the condenser (c).

wood, Mouchot boiled the alcohol by means of solar energy. The glass-covered copper cauldron served as the boiler in which wine was heated to a vapor. The vapor then cooled and collected in a conventional receiver. In his first experiment, Mouchot filled the cauldron with two quarts of wine. A few hours later he had the pleasure of drinking the first brandy ever distilled by the heat of the sun, remarking that it had a "most agreeable flavor."

Mouchot's solar pump was similar to the basic design of the solar oven and still. A tall, hollow copper cauldron surrounded by two glass covers was soldered on top of a short tank filled with water. A cylindrical reflector concentrated the rays of the sun on the cauldron, rapidly heating the air inside. The expanding air exerted pressure on the water in the container below. Within 20 minutes enough pressure had built up to shoot a jet of water through a nozzle attached to the container, producing a spray ten feet long that lasted over half an hour. During succeeding attempts, it pumped a continuous stream of water 20 feet. Mouchot patented his first primitive sun machine on March 4, 1861.

The First Solar Engines

Despite these successes, Mouchot had not yet attained his main goal: to drive a steam engine with sun power. The large volume of water inside the copper cauldron took a long time to boil, and the device produced steam too slowly to drive an industrial motor. Mouchot therefore substituted a one-inch diameter copper tube for the cauldron so that the smaller volume of water in the tube would heat much faster, generating steam more quickly. To collect the steam Mouchot soldered a metal tank to the top of the tube. The solar reflector consisted of a parabolic trough-shaped mirror that faced south and was tilted to receive maximum solar exposure.

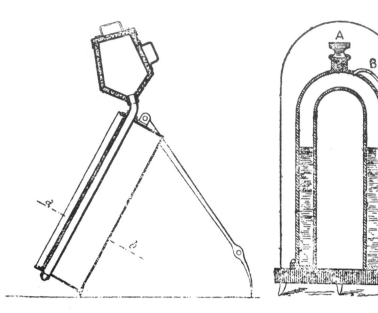

Right: Boiler in the first successful solar-powered steam engine, 1866. A parabolic trough reflector concetrated the sun's rays onto a copper tube, generating steam that collected in a metal tank above.

Far right: Cross-section of Mouchot's improved, double-cauldron boiler, from La Chaleur Solaire.

Excitedly, Mouchot reported what happened when he connected this boiler to a specially designed engine:

> In the month of June, 1866, I saw it function marvelously after an hour of exposure to the sun. Its success exceeded our expectations, because the same solar receptor [i.e., reflector and boiler] was sufficient to run a second machine, which was much larger than the first.

Mouchot had invented the first steam engine to run on energy from the sun! He presented it to Napoleon III, who received it favorably.

Over the next three years Mouchot continued to refine his solar motor. To increase the boiler's steam-generating capacity so that it could run a large industrial-sized machine, he replaced the copper tube with two bell-shaped copper cauldrons, one inside the other. The double cauldron was sheathed in glass. The space between the copper shells held a somewhat greater volume of water than the previous model, but the layer of water was thin enough to heat rapidly.

The French government gave Mouchot the financial backing to construct an industrial-scale solar engine along these lines. For this project he built a seven-foot long cylindrical boiler based on the double-cauldron design. It had a tall trough-shaped reflector that faced south. Mouchot added a clock mechanism that automatically moved this device from east to west to follow the daily course of the sun, which had previously been done manually.

On the whole, the boiler's performance satisfied Mouchot: it vaporized water into steam at a pressure of 45 pounds per square inch. But he recognized that the mirror needed improvement. First, because the mirror's tilt could not be adjusted for seasonal shifts in the sun's path, it did not reflect an optimum amount of sunlight throughout the year. Second, the mirror concentrated sunlight onto only one side of the boiler—the opposite side remained cooler, lowering the overall efficiency of the machine. Third, the reflector was constructed of silver plates and wood. Mouchot

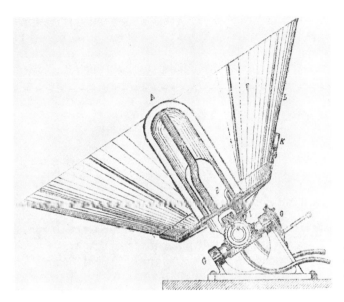

The Tours solar engine, 1874.
With the boiler (A) placed
along its axis, this conical
reflector pivoted to follow the
sun's daily and seasonal
motions.

was worried that a mirror large enough to run industrial equipment would be too
heavy for the clock mechanism to move.

An Improved Sun-Powered Motor

International conflict halted further experimentation. Napoleon III declared war
on Prussia in July of 1870; soon afterwards the enemy marched into Paris and the
French suffered an ignominious defeat. One of the victims of the conflict was
Mouchot's solar reflector—disappearing, as he put it, "in the midst of our disasters." After the government collapsed and Mouchot's funds were cut off, he had to
seek other sources of support. But Mouchot had a hard time obtaining backing,
which he attributed partly to the "bias and specious objections that engineers
pronounced on a question too foreign to their own studies for them to judge."

Eventually he took the advice of a colleague and approached the regional
government of Endre-et-Loire, the wine-producing district in which he lived. The
solar device he showed them "so pleased the general counsel that they gave me
1,500 francs on the spot," remarked a happy Mouchot, "making available to me the
means for completing the construction of a large solar receptor capable of distilling
alcohol and producing a sufficient amount of power for mechanical applications."

With customary alacrity, Mouchot finished his new solar machine by 1874. He
put it on public display in Tours, the district capital, where the journalist Leon
Simonin saw it and reported:

> The traveller who visits the library of Tours sees in the courtyard in front a
> strange-looking apparatus. Imagine an immense truncated cone, a mammoth
> lamp-shade, with its concavity directed skyward.

This "strange-looking apparatus" was Mouchot's redesigned solar reflector. Made
of copper sheets coated with burnished silver, the inverted cone-shaped mirror

measured 8½ feet in diameter at its mouth and had a total reflecting surface of 56 square feet. The mirror's huge size required it to be constructed in sections, the technique developed by Hoesen a century before. Mouchot borrowed the idea for the reflector's novel shape from a fellow mathematician, Dupuy. It had the advantage of correcting one of the major defects of previous models: its surface could reflect sunlight at right angles to all sides of the boiler, which was located along the axis of the cone. Simonin described the boiler's configuration:

> On the small base of the truncated cone rests a copper cylinder; blackened on the outside, its vertical axis is identical with that of the cone. This cylinder, surrounded as it were by a great collar, terminates in a hemispherical cap, so that it looks like an enormous thimble. . . . This curious apparatus is nothing else but a solar receiver . . . or in other words, a boiler in which water is made to boil by the rays of the sun.

The machine could generate enough steam to drive a ½-horsepower engine at 80 strokes per minute. When Mouchot put it on display, the reaction was one of amazement—a motor that ran without fuel, on nothing more than sunbeams! On one occasion the crowd became somewhat apprehensive when the steam pressure rose to over 75 pounds per square inch. One nervous bystander exclaimed, "It would have been dangerous to have proceeded further, as the whole apparatus might have been blown to pieces." People were also impressed that the boiler could operate a commercial distiller capable of vaporizing five gallons of wine per minute.

The success of the Tours machine bolstered Mouchot's belief in sun power. But he also became aware of the limits of its practical application in a country like France. First of all, he realized that "sun machines would take up too much space in our cities and therefore could not be profitably used." The Tours motor occupied an area 20 ft by 20 ft and produced only one-half horsepower. Two hundred such machines would be needed to drive a typical 100-horsepower industrial motor. If 200 solar machines were arranged in four lines, with enough space between each device to prevent them from casting shadows on one another, a total of 100,000 square feet would be needed.

Sun Power for the Colonies

An additional impediment to the commercial use of solar engines in France was the problem of intermittent sunshine—especially during winter. But Mouchot believed that France's sun-baked colonies in North Africa and Asia, many of which had recently been conquered and were just being opened up to French exploration and settlement, offered unlimited possibilities. For example,

> In torrid zones such as Cochin-China [South Vietnam], the matter of hygiene comes to the fore. In Saigon, water has to be boiled to be made potable. What a savings in fuel one could realize using a [solar still] in the ardent heat of those climates!

The combination of nearly constant sunshine and abundant open space convinced Mouchot of the commercial viability of solar power in these colonies. Hence, he sought financial aid to develop his machines for such areas.

Mouchot's opportunity came a year after the demonstration at Tours, when he

*Mouchot developed
this portable solar
oven for French
troops in Africa.*

had the good fortune to meet the Baron of Wattville. The baron became a strong
advocate of solar power, and through his influence the French Ministry of Public
Education agreed to pay for a scientific expedition to the colony of Algeria so that
Mouchot could determine whether solar cooking, distilling, and pumping would be
practical there. The French government wanted to aid the new wave of colonial
settlement in Algeria that followed their suppression of the insurrection of 1871.
The subsequent confiscation of tribal lands and commercial holdings made it easier
for Europeans to acquire property, but lack of indigenous fuel supplies hindered
economic development. Algeria had to import all her fuel from Europe, 85 percent
of it from England in the form of relatively expensive coal. Lack of a railroad system
in the colony drove the cost of coal up even further—especially in remote districts
where prices were nearly ten times higher than in more accessible regions. The
French government hoped that solar power could be a great boon to Algeria's
economy.

Mouchot arrived in Algiers on March 6, 1877, and immediately set to work
testing an improved solar oven. ''It seems possible without any great cost to
provide our soldiers in Africa with a small and simple portable [solar] stove,
requiring no fuel for the cooking of food,'' he wrote. ''It would be a big help in the
sands of the desert as well as the snows of the Atlas [Mountains].''

The oven Mouchot designed had a truncated conical reflector, like the one at
Tours, with a glass-enclosed cylindrical metal pot—serving as the boiler—sitting at
the focus of the reflector. The entire apparatus weighed only 30 to 40 pounds and
could be collapsed and packed into a 20 in. by 20 in. box. Before a group of
appreciative spectators Mouchot baked a pound of bread in 45 minutes, over two
pounds of potatoes in one hour, a beef stew in three, and a perfect roast—''whose
juices fell to the bottom'' of the pot—in less than half an hour.

By removing the pot and replacing it with bottles of freshly processed wine, the oven became a pasteurizer. The sun heated the bottles, killing any traces of bacteria that might later multiply when the wine was being shipped over long distances. Mouchot envisioned that Algeria would—

> Be able to ask from the beautiful sun not only to care for the ripening of her vineyards, but also to improve her wines and make them transportable, which would be for her a new source of prosperity.

Mouchot also tested a solar still, similar in size and design to the solar machine exhibited at Tours. Mouchot wrote that the brandy he made from wine was "the subject of astonishment . . . [for] it is undeniable that the alcohol comes out of the solar still bold [and] agreeable to the taste, and with an appropriate wine it offers the savor and bouquet of an aged '*eau-de-vie*'." The device could distill fresh water and salt water as well.

Mouchot travelled from the Sahara to the Mediterranean to test the feasibility of using solar water pumps for irrigation. Irrigation was crucial to agricultural development in Algeria, and coal-powered pumps were too expensive. Mouchot experimented with a solar pump similar to the Tours machine, and found that it worked more reliably in Algeria's sunny weather than in France's variable climate.

Following a year of testing, Mouchot presented his findings to the authorities in Algiers. They were so impressed that they awarded him 5,000 francs to construct "the largest mirror ever built in the world" for a huge sun machine that would represent Algeria in the upcoming Universal Exposition in Paris. Afterwards, it was to be shipped back to Africa and used commercially.

With the help of his assistant, Abel Pifre, Mouchot completed this new solar machine in September 1878. At its widest point the cone-shaped mirror measured twice the diameter of the device shown at Tours the previous year, and its total reflecting surface was four times greater. The boiler had an unusual design: a group of long vertical tubes were fastened side by side to form a circular column at the focus of the reflector.

Mouchot's solar machine astounded exposition visitors—pumping over 500 gallons of water per hour, distilling alcohol, and cooking food. But the most remarkable demonstration occurred on September 22, as Mouchot recounted:

> Under a slightly veiled but continually shining sun, I was able to raise the pressure in the boiler to 91 pounds . . . [and] in spite of the seeming paradox of the statement, [it was] possible to utilize the rays of the sun to make ice.

He had connected the solar motor to a heat-powered refrigeration device invented by Ferdinand Carré in the 1850's. Mouchot saw an important future for solar refrigerators in hot climates, where sun-generated ice would help prevent perishables from spoiling.

Solar Electricity

The following year Mouchot returned to Algeria to resume his research. He spent much of his time trying to resolve a difficult question: how could solar heat be

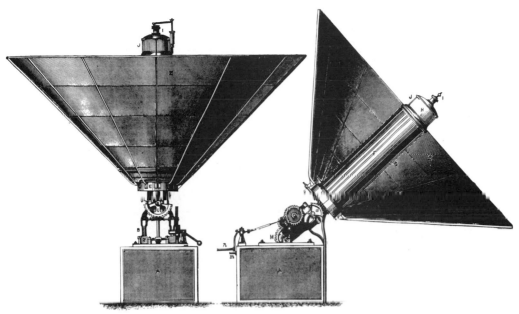

Two views of the solar motor built by Mouchot for the Paris Exposition. This giant machine pumped 500 gallons of water per hour, and even powered an ice-maker.

stored so that sun machines would be able to work during cloudy weather or at night? A colleague suggested using heat-absorbing materials capable of withstanding the high temperatures produced by a solar reflector. By placing these solar-heated substances in an insulated container, solar heat would be retained for later use.

But Mouchot discovered what he thought was a better alternative. If solar energy were used to break down water into hydrogen and oxygen, the gases could be stored in separate cylinders. When heat was needed, the chemical reaction resulting from recombining the two would produce very high temperatures. Or, the gases could be used separately—the hydrogen as fuel and the oxygen for industrial purposes. As for the method of separating water into its components, Mouchot decided to try "an instrument already in excellent condition . . . the thermoelectric device." The principle behind its operation was simple: when two different metals such as copper and iron are soldered together and heat is applied to the juncture between them, an electrical current results. Mouchot planned to heat a hundred such metallic couplings with a solar reflector, and thereby generate enough electricity to change water into its constituents.

By 1879 Mouchot had "already made a few experiments which bode well for this procedure. . . . Some very primitive devices have given me significant amounts of electricity," he wrote. Mouchot had great expectations, and hoped to decompose water and produce "a reserve of fuel that would be as precious as it is abundant." But for all his efforts, he could not compete with the more efficient methods of electrical generation rapidly being perfected about the same time.

In 1880, Mouchot returned to his mathematical studies. His assistant, Abel Pifre,

Abel Pifre's solar-powered printing press, 1880. Exhibiting it at the Gardens of the Tuileries, he printed 500 copies of the Solar Journal.

took over the solar research. Pifre built several sun motors and conducted public demonstrations to gain support for solar power. At the Gardens of the Tuileries in Paris, he exhibited a solar generator that drove a press which printed 500 copies of the *Journal Soleil* [Solar Journal].

But the time wasn't right for solar energy in France. The advent of better coal-mining techniques and an improved railroad system (most of France's coal lay at her borders) increased coal production and reduced fuel prices. In 1881 the government took one final look at the potential for commercial use of solar energy: it sponsored a year-long test of two solar motors—one designed by Mouchot and the other by Pifre. The report concluded:

> In our temperate climate [France] the sun does not shine continuously enough to be able to use these devices practically. In very hot and dry climates, the possibility of their use depends on the difficulty of obtaining fuel and the cost and ease of transporting these solar devices.

Furthermore, the cost of constructing silver-plated mirrors and keeping them highly polished proved economically prohibitive for most uses. During ensuing years, however, the French Foreign Legion made some use of solar ovens in Africa. In remote areas of Algeria people also used solar stills to obtain water, as one Paris correspondent reported:

> One of the great services that we owe to Mouchot's appliances is the distillation of brine water heavily charged with magnesium salt, which is abundant in the African desert. [His still] is a great benefit to settlers and explorers.

Although Mouchot did not succeed in bringing France into the "Sun Age," his pioneering efforts crossed the threshold between scientific experimentation and the practical development of a revolutionary technology. He demonstrated a great variety of ways that solar energy could be used to benefit humankind and laid the foundations for future solar development.

John Ericsson's final sun motor, which he hoped would provide cheap energy for the world's fuel-short regions.

Chapter 7
Two American Pioneers

Eight years after Augustin Mouchot began his first experiments, an American engineer named John Ericsson also voiced the hope that someday solar energy would fuel the machines of the Industrial Age. Born in Sweden in 1803, Ericsson emigrated to America in 1839. By that time he was already well known for such inventions as the screw propeller, which made steam navigation practical. He later achieved widespread recognition for designing the ironclad battleship *The Monitor*, which defeated the Confederate iron ship *The Merrimack* off the Virginia coast on March 9, 1862. *The Merrimack* had wrought havoc on the wooden ships of the Union Navy; if not for *The Monitor's* decisive victory, the Confederacy would have controlled the seas—perhaps changing the course of the American Civil War. Six years later, Ericsson set his sights on a more peaceful goal: to bring his dream of sun power to fruition.

Like Mouchot, Ericsson was deeply concerned about the rapid consumption of coal. In a paper written in 1868, he contended that only the development of solar power would avert an eventual global fuel crisis. He pointed to an example from the previous century—the evaporation of sea water to produce salt. This was accomplished by solar heat rather than coal, saving some 100 million tons of fuel. How much greater savings could be realized if solar motors were developed! Ericsson prophesied that with commercially viable solar engines the sunnier parts of the world could become the source of virtually limitless power:

> A great portion of our planet enjoys perpetual sunshine. The field therefore awaiting the application of the solar engine is almost beyond computation while the source of its power is boundless. Who can foresee what influence an inexhaustible motive power will exercise on civilization and the capability of the earth to supply the wants of our race?

Thus he declared that he would dedicate the balance of his life to making solar engines an economical alternative.

Ericsson completed the construction of a solar-powered steam engine in 1870. He erroneously claimed it as the world's first, brusquely dismissing Mouchot's solar-driven motor built four years earlier as a "mere toy." However, Ericsson's own invention—which he ironically intended to present to the French Academy—bore a striking resemblance to Mouchot's. Both had three components: a concentrating mirror, a boiler, and a steam engine. Ericsson had ruled out the feasibility of a hot-box collector on grounds similar to Mouchot's. A hot box, it appeared, could not produce sufficient steam to drive an engine unless it was enormous. Unlike Mouchot, though, Ericsson decided to abandon the concept of a glass heat trap entirely. Instead, bare metal tubes serving as the boiler were placed at the focus of a series of parabolic, trough-shaped solar reflectors. This collector generated enough steam to run a small conventional engine.

Little more is known about this sun motor. Ericsson was emphatic about keeping its details secret: "Drawings and descriptions of the mechanism . . . will not be presented, nor will the form of the generator . . . be delineated or described," he stated. He had had some unfortunate experiences with "enterprising persons" who filed patents on slightly modified versions of his inventions, preventing him from making similar improvements on his own devices.

Right: Portrait of Captain John Ericsson, a Swedish-American inventor and solar pioneer, 1876.

Below: Steam engine used with Ericsson's first solar motor, 1870.

A Solar Hot-Air Engine

In 1872, Ericsson tried a very different approach for his next solar motor: it was powered by solar-heated air, rather than steam. This unique invention was a logical next step, for he had spent much of his early career developing the hot-air external combustion engine. This type of motor worked on the following principle: heat externally applied to a cylinder caused air inside the cylinder to expand and push down a piston; inrushing cold air pushed the piston up again, and the cycle repeated itself. Normally wood or coal was used to provide the heat. Ericsson modified the process by placing the cylinder upright along the axis of a curved, dish-shaped reflector. The mirror concentrated the rays of the sun onto the top of the cylinder, heating the air inside. For this motor, as for all of Ericsson's other solar engines, the reflector had to be moved manually to stay aligned with the sun.

Elated at the hot-air engine's performance, Ericsson wrote a letter to his close friend and associate, Harry Delameter:

> The world moves—I have this day seen a machine actuated by solar heat applied directly to atmospheric air. In less than two minutes after turning the reflector toward the sun the engine was in operation. . . . As a working model I claim that it has never been equalled; while on account of its operating by a direct application of the sun's rays it marks an era in the world's mechanical history.

But three years and five more experimental engines later, Ericsson's enthusiasm had been tempered. He realized that "although the heat is obtained for nothing, so extensive, costly, and complex is the concentrating apparatus" that engines powered by solar energy were actually more expensive than similar coal-fueled motors.

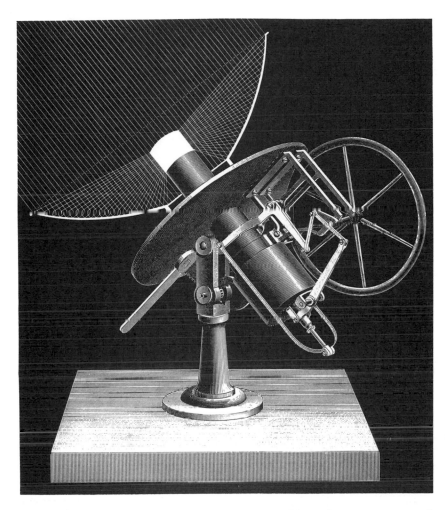

Engraving of the world's first solar hot-air engine. The curved mirror concentrated sunlight upon a piston at its focus.

The concentrating mirror, usually made of silver sheets or some other silvered metal, cost too much to manufacture and keep from tarnishing.

Another obstacle to the commercial viability of solar motors confronted him, just as it had hindered Mouchot: the use of solar energy was restricted to daylight hours in areas enjoying almost constant sun. Ericsson felt he could not "recommend the erection of solar engines in places where there is not steady sunshine until proper means shall have been devised for storing up the radiant energy in such a manner that regular power may be obtained from irregular solar radiation." But his decades of searching for a way to store solar heat were of no avail.

An Inexpensive Solar Reflector

Ericsson had better luck in developing a way to cut down the cost of the solar reflector. He replaced the silver-coated mirror with window glass silvered on its underside. Because this silver finish was not exposed to the elements and was further protected by a special coating, the mirror did not tarnish. "Hence, it will

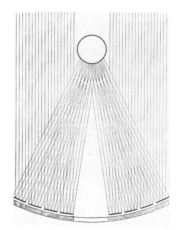 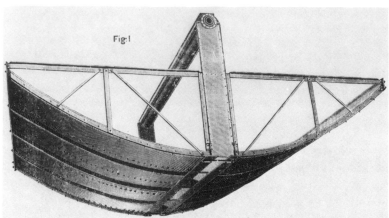

The lightweight parabolic-trough reflector developed by Ericsson in 1884.

only be necessary to remove dust from the mirror" to maintain its reflectivity, Ericsson pointed out, "an operation readily performed by feather dusters." The reflector could be purchased for less than 60 cents per square foot—much cheaper than a polished silver mirror—and the cost of upkeep was minimal.

The design of this reflector also marked a departure from Ericsson's previous models. Individual silvered glass mirrors were attached to a metal frame to form a parabolic trough-shaped reflector. A tubular boiler was mounted along the mirror's focus.

In 1884 Ericsson unveiled his new reflector design, claiming that its cost would not exceed that of a conventional steam boiler. A number of California farmers contacted him about buying the motor, hoping to obtain a cheap power source for their irrigation pumps. But Ericsson's announcement had been premature—some problems remained to be worked out. "Consequently," he explained, "no contracts for building sun motors could then be entered into." After four years of refining the model, Ericsson felt he had made enough progress to offer the solar pump for sale. In the fall of 1888 he confidently stated:

> The new motor being thus perfected, and first class manufacturing establishments ready to manufacture such machines, owners of the sun-burnt lands on the Pacific Coast may now with propriety reconsider their grand scheme of irrigation by means of sun power.

But only seven months later, before any of these plans had materialized, Ericsson died at age 86. According to his obituary in *Science*, Ericsson's interest in solar motors obsessed him even on his death bed:

> He continued to labor at his sun motors until within two weeks of his death. As he saw his end approaching, he expressed regret only because he could not live to give this invention to the world in completed form. It occupied his thoughts to the last hour.

And due to Ericsson's penchant for secrecy, the details of what he called his "perfected solar motor" followed him to his grave.

Interest in Solar Energy Grows

Ericsson's vision of harnessing solar power remained very much alive in the minds of other American engineers and scientists. Many had warned of an impending fuel crisis—warnings largely ignored by the public. But the devastating effects of a series of coal strikes around the turn of the century, culminating in a massive strike in the winter of 1902, threw "a new and lurid light on [these predictions] . . . for many a home has been fireless and many a factory has closed its doors," according to *Harper's Weekly*. Charles Pope, author of *Solar Heat*, one of the first books on solar energy, agreed: "The year of 1902 has added an awful chapter to the history of our need of a new source of heat and power," he wrote.

To Pope and others, solar energy appeared to be the most promising alternative to coal for powering machines such as water pumps. In an article reprinted in the *Smithsonian Annual Report* of 1901, Robert Thurston, a renowned engineer, reviewed the pros and cons of tidal, wind, and solar power as replacements for coal. American farmers and ranchers commonly had small windmills to pump water for their houses or livestock, but windmills were inadequate for large-scale irrigation. As Thurston noted, they were "unreliable for steady work":

> Rising to a gale and falling to a calm, alternately, the portion of time during which this power is actually available is small, and, still worse, its available periods are as likely to come at unsuitable hours and seasons as when wanted. . . . [Therefore] it does not seem likely that this particular problem will be successfully solved even under the stimulus of vanishing fuel supplies.

Tidal power, too, did not appear to be a viable substitute for coal at the time.

However, Thurston observed that solar energy was then attracting great attention:

> Engineers and men of science are studying the art of harness[ing] the direct rays of the sun, and the solar engine is exciting special interest. . . . Probably at no time in the past has this matter assumed importance to so many thoughtful and intelligent men or excited so much general interest.

Indeed, by the turn of the century numerous solar inventors and entrepreneurs crowded the field. At least 22 patents had been filed for solar motors. But few of these patents ever led to workable solar engines. As Charles Pope wrote, "There comes a strong suspicion that the patent office admitted essayists, rather than inventors to their lists; and that these men were not actually makers of machines which did what they claimed." Mixed in this whirlwind of solar enthusiasm were charlatans and dreamers as well as some of the most talented inventors of the time.

Eneas' First Solar Pump

One of the more successful leaders of the turn-of-the-century solar movement was Aubrey Eneas. An English inventor and engineer residing in Massachusetts, he began studying sun power in 1892. With the aid of investors, he founded the Solar Motor Company of Boston to design and manufacture commercial solar engines. Like Ericsson, Eneas was especially interested in marketing solar-powered irrigation pumps in the American Southwest where conventional fuels like wood and coal were scarce and expensive. This part of the country seemed ideal for solar power because 75 percent of the days were sunny.

Not much detailed information on solar motors had been published up to that time. "Though I have had technical training and considerable engineering experience," Eneas wrote, "I find so much that is new and confusing, and so little data of a practical nature." Probably all that he had at his disposal were the meager accounts of Ericsson's work printed in the scientific periodical *Nature*.

In 1898 Eneas built his first solar motor, almost an exact replica of Ericsson's 1884 model—with silvered glass mirrors forming a trough-shaped reflector, and a bare metal cylinder at its focus. This engine was 1½ times larger than Ericsson's, and preliminary experiments led him to believe that it was—

> A step in advance of what has already been accomplished, of practical value to our great arid west, in affording to this district an irrigation engine that requires but little care and no fuel.

But Eneas had fallen prey to the same premature optimism as Ericsson. Although plans were made to sell the motor for $1,500, more extensive tests proved disappointing. After a full summer of operation, Eneas concluded that the reflector had several serious drawbacks. According to engineering knowledge of that day, generating sufficient steam to run a fair-sized engine efficiently required boiler temperatures exceeding 1,000°F. But the parabolic mirror Eneas had been using could not produce such a high temperature, and increasing the reflector's size would make it cumbersome to support and difficult to maneuver. Furthermore, the reflector's shape was also inefficient. As Mouchot had discovered earlier, the reflector concentrated sunlight only onto the side of the boiler it faced; the opposite side of the boiler merely received diffuse solar rays, and more heat actually escaped from this side than was collected there.

Back to the Conical Reflector

Eneas decided to abandon his first design and build a reflector shaped like an inverted, truncated cone with a boiler standing upright along its axis—much the same as Mouchot had done 30 years earlier. Perhaps he had read of Mouchot's work in an abridged account published by Ericsson in 1876.

In subsequent experiments, Eneas made some important observations about the maximum efficiency of the conical reflector. Because the mouth of the reflector has the widest diameter, the mirror surface is greater at this end and more sunlight is concentrated onto the upper end of the boiler. Conversely, less sunlight is concentrated onto the lower end of the boiler near the bottom of the reflector. In fact, the lowest part of the boiler inside Mouchot's reflector received less heat from the sun's rays than it lost to the outside air. For maximum steam production, Eneas calculated that the smallest diameter of the mirror should be at least eight times the diameter of the boiler and the larger diameter should be at least 32 feet wide—nearly 2½ times wider than Mouchot's reflector.

Accordingly, Eneas built a truncated-cone reflector measuring 33½ feet in diameter at its mouth and 15 feet in diameter at its bottom. The ratio of reflecting

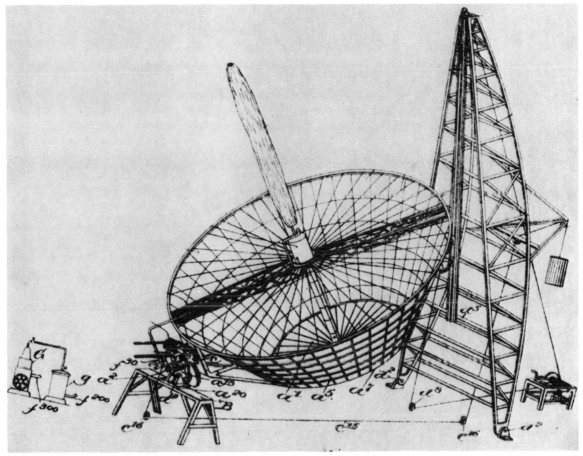

Patent drawing–Aubrey Eneas' second solar motor, 1899. To follow the sun's seasonal motion, this truncated-cone mirror was raised or lowered along a track in the vertical tower behind it.

surface to boiler area was 25 to 1, so that nearly twice as much sunlight was concentrated on the boiler as in Eneas' previous model. The reflector consisted of over 1,800 small silvered glass mirrors. The boiler resembled one of Mouchot's later designs: two concentric metal shells held 100 gallons of water between them. Although Eneas first covered the boiler with a glass jacket, the final version of the motor had a bare metal boiler.

Eneas made a lightweight but strong metal tower to support the 8,000-pound machine so it could be angled to receive optimum sunlight. The high point of the reflector slid up or down a track in the vertical tower, directing the mouth toward the high summer or low winter sun. To follow the daily motion of the sun, a clock mechanism automatically rotated the mirror from east to west.

By 1899 Eneas had finished building this solar engine. But the New England sunshine was insufficient to produce satisfactory test results. He decided to ship the motor to Denver, Colorado, where "the direct rays of the sun are intensely hot" due to the high altitude. The Denver tests convinced Eneas and his backers that

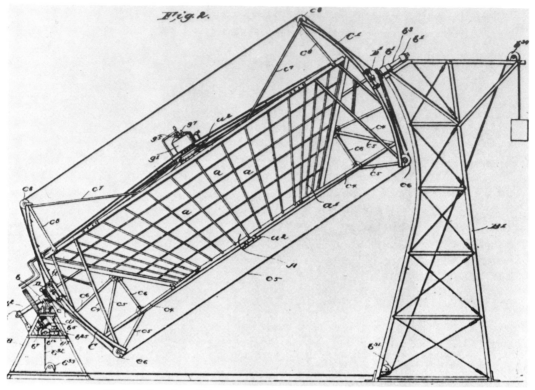

Patent drawing of the improved reflector mounting in Eneas' sun motor. In this version, the reflector could rock back and forth along two curved tracks as the sun's path changed seasonally.

they were on the right track, although they discovered one major flaw. The machine was unevenly balanced—it required great force to raise the massive device and adjust its angle to the sun's seasonal motion. Eneas solved this problem by mounting the mirror in a cradle-like frame that distributed the mirror's weight evenly. At opposite ends of the reflector, he installed arched runners that allowed it to rock back and forth. The tilt of the mirror could then be changed effortlessly. In this way "the machine will at all times be in proper equilibrium," Eneas declared, and it will be "possible to operate it with minor power and under all conditions of wind and weather."

Solar Power on the Ostrich Farm

Satisfied with this alteration, Eneas moved the motor to the sunny climate of southern California in 1901 for extensive tests and public demonstrations.He arranged to put the machine on display at the only ostrich farm in America, located in Pasadena. It belonged to Edwin Cawston, a fellow Englishman, who raised these gangly birds to supply the fashion industry with their popular plumes. The ostrich farm had become a national tourist attraction, and Eneas saw it as the perfect place for his invention to receive a great deal of public exposure. On his part, Cawston thought the solar motor would be an extra attention-getter for his farm. He advertised on handbills:

NO EXTRA CHARGE TO SEE THE SOLAR MOTOR—The only machine of its kind in the world in daily operation. Fifteen horsepower engine worked by the heat of the sun.

As Cawston and Eneas hoped, thousands saw the sun machine at the ranch. Anna Laura Myers, who visited the ostrich farm many times during her childhood, recalled: "It was one of southern California's wonders. A strange thing, just as strange as an ostrich. People would just look at it and wonder about it."

As news of the sun motor spread, over a dozen popular and scientific publications sent reporters to cover the story of Eneas and his machine. "Sun power is now at hand!" many enthusiastically proclaimed. Even the more conservative articles indicated that sun power was at last "within the domain of what may be called practical science." F.B. Millard, one of the journalists who came out to the Cawston ranch, witnessed the solar motor operate a pump capable of irrigating 300 acres of citrus trees by drawing 1,400 gallons of water per minute from a reservoir 16 feet deep. Millard was convinced that "solar motors will before long be seen all over the desert as thick as windmills in Holland, and that they will make the desert blossom as the rose—a phrase that literally represents the possibilities of the machine." He also saw the sun motor as a source of inexpensive power for domestic purposes, which meant "cheap homes in the arid regions . . . homes for millions of men where there are now only hundreds."

On a typical day the machine began operating 1½ hours after sunrise and continued until an hour after sunset. To start the motor the attendant turned the reflector toward the rising sun. As the rays struck the mirror, the boiler heated up. Mr. Millard described the spectacle:

At first the morning dew is seen slowly to ascend in a wreath of vapour from the gigantic mouth. Then the bright glasses glitter in the morning sun, and the heatlines begin to quiver inside the circle, the greatest commotion being about the long, black boiler, which, as the intensity of the focused rays increases, begins to glisten so that in any photograph taken of the machine, the boiler is shown almost pure white. Within an hour of the turning of the crank and getting the focus there is a jet of steam from the escape valve. The engineer moves the throttle, there is a succession of hisses from the boiler, a "clank-clankety-clank!" and the sun is drawing water in a way which he little dreamed a few months ago.

The reflector automatically followed the sun's course throughout the day. Steam pressure as well as water supply to the boiler were also self-regulated, and the motor even oiled itself. So little attention did the engine require that, according to Mr. Millard, the operator had enough time off to "hoe his garden, or read his novel, or eat oranges, or go to sleep."

Another reporter, impressed by the almost incredible heat reflected by the mirror, wrote:

Should a man climb upon the [reflecting] disk and cross it, he would literally be burned to a crisp in a few seconds. And a pole of wood thrust into the magic circle flames up like a match.

1

2

3

4

1 *The Pasadena sun motor in operation. Notice the glistening white boiler and the steam escaping from a relief valve.*

2 *A workman nimbly avoiding the focus of the giant mirror. There the concentrated sunlight generates visible "hot spots" on the boiler behind him.*

3 *Handbill advertising Cawston's ostrich farm and Eneas' solar motor, 1901.*

4 *The improved solar motor on display at Cawston's ostrich farm.*

Marketing the Solar Motor

Buoyed by the glowing publicity and two years of successful operation of the solar pump at the ostrich farm, Eneas felt ready to offer his invention for sale. In 1903 he incorporated the Solar Motor Company in California and opened offices in Los Angeles' prestigious Bradbury Building. The complete machine—reflector, boiler, steam engine, and pump—sold for $2,160.

Eneas made his first sale to Dr. A. J. Chandler, who owned a large tract of land just outside of Mesa, Arizona, 35 miles southeast of Phoenix. By the summer of 1903 the company had set up the machine on Chandler's land. Unfortunately, after a short period of operation the frame supporting the mirror weakened, and under the force of a windstorm the reflector toppled. It was badly damaged. As Pete Estrada, who was twelve at the time, remembered: "The metal frame that held it up was still intact, but the top of it, the mirror, was down all over the ground, broken, every bit of it."

But the accident did not discourage interest in sun power. If California needed an economical method of irrigation, Arizona required one desperately. The following year the Santa Fe Railroad, seeking to encourage the development of the American Southwest, persuaded the Solar Motor Company to put a sun machine on public display. By March the firm had installed one in Tempe, a small town eight miles southeast of Phoenix. The idea of such a large and unusual contraption being set up near the local high school did not rest well with Tempe's provincial residents, many of whom had not yet seen a car. But the demonstration proceeded as planned, opening on March 21, 1904.

In July, John May purchased the motor for his farm located in hot and desolate Sulphur Springs Valley, not far from Wilcox, Arizona. With the help of Bert Parker, a local mechanic, construction was completed by September. On the opening day Parker wore his Sunday best to launch the machine's maiden run. Before a crowd of people from all over the county, believers as well as skeptics, Parker stepped up to the platform. He turned the huge mirror toward the rising sun, and soon the steam pressure in the boiler shot up to 110 pounds. As he gradually opened the steam valve until the boiler was going full blast, the motor—wrote one journalist—looked "like a bucking horse that had to be tamed." It took a few more minutes to adjust the machine properly and get it running smoothly. A powerful four-inch spray of water jetted out of the pump's nozzle, and the water broke through a makeshift dam built to contain the flow. Even the doubters were converted, and the spectators left the farm confident that a new era had begun in the Southwest. One account of the occasion likened it to that day at Kitty Hawk three years earlier when the Wright brothers proved that humanity could loosen its earthly bonds.

On sunny days the machine steadily pumped 1,500 gallons of water per minute, enabling Mr. May to grow "corn that would do Iowa credit; watermelons that were so big and luscious that they could have taken a prize at any man's fair," as one reporter observed. "It was a sight to see in arid Arizona."

But the sun motor's performance was not the only point a potential customer had to consider. The destruction of the pump at the May ranch by a hailstorm some months later added to growing skepticism about the machine's ability to withstand the harsh weather of the Southwest. Unscrupulous rivals in the solar business also tainted Eneas' reputation. *Harper's Weekly* referred to such people as "wild-eyed wizards with companies to promote." But the major obstacle to marketing Eneas' engines was their cost—$2,500 plus $500 for installation. This came to $196 per horsepower—two to five times the cost of a conventional steam plant. Even though no fuel expenses were incurred after the initial investment, the high purchase price deterred buyers. In the following years, the commercial prospects for Eneas' solar machines dimmed. "My . . . experience with large reflectors has convinced me," he wrote, "that even where the greatest efficiency is obtained, the cost of construction, even on an extensive scale, is too great to permit of their use in a commercial way, except in a few instances."

While Eneas felt he had reached a dead end with reflectors, others began to consider other, more economical ways to produce solar power for industry and agriculture. Low-temperature solar engines, requiring neither expensive mirrors nor the high heat they generated, appeared to be one solution.

A sun motor at John May's farm helped make the Arizona desert bloom.

Charles Tellier's solar pump, built onto his shop near Paris during the early 1880's. Instead of water, this engine used ammonia as the working fluid.

Chapter 8
Low
Temperature
Solar Motors

The solar engines developed by Mouchot, Ericsson, and Eneas all relied on the use of reflectors to concentrate the sun's rays. In their eager attempts to build a commercially viable engine, these early pioneers assumed that some concentration of solar energy would be needed to achieve the high steam temperatures they thought were essential. Because conventional wisdom taught that the higher the steam temperatures the more efficiently an engine would run, they sought to produce steam at temperatures in excess of 1,000°F.

Unfortunately, this emphasis on concentrating collectors and high temperatures led to a number of critical drawbacks. As de Saussure had demonstrated a century earlier, high temperatures inside a collector inevitably caused large heat losses. Hence, even though high temperatures meant greater engine efficiency, solar collection efficiency dropped substantially—bringing down the overall efficiency of converting solar energy into mechanical power.

There were other drawbacks, too. The reflectors used by Mouchot, Ericsson, and Eneas were complex and expensive pieces of equipment. Once installed they were vulnerable to high winds and inclement weather. To make matters worse, they always had to face the sun, which required either a full-time attendant or a delicate mechanism to move the reflectors automatically. And when there was no direct sunlight on hazy or cloudy days, these concentrating collectors could not function at all.

Solar reflectors would be unnecessary in an engine designed to operate at lower temperatures. Simple, inexpensive hot boxes, which Mouchot and Ericsson had rejected, or even bare metal plates could be used instead, to eliminate most of the problems. Because they did not reach such high temperatures, these collectors would not lose as much heat. Furthermore, the collecting surface would not have to follow the sun's motion. Such devices could absorb even diffuse sunlight during hazy or cloudy weather. Better solar collection efficiency and lower construction costs appeared to outweigh the loss (due to lower operating temperatures) in engine efficiency. Late in the nineteenth century a number of inventors began to realize these advantages, and started to develop low-temperature solar engines.

A Low-Temperature Solar Pump

Charles Tellier, a French engineer, was the first person in modern times to develop low-temperature solar collectors to drive machines. Whereas conventional engines used pressurized steam, Tellier's devices used pressurized vapor from certain liquids having boiling temperatures well below that of water. Ammonia hydrate, for example, will boil at −28°F; sulphur dioxide will boil at 14°F. These substances vaporize rapidly when exposed to higher temperatures.

Tellier discovered the unusual properties of low boiling-point liquids during his research in cold food storage. Called the "Father of Refrigeration," Tellier radically transformed international trade by enabling the ship *Frigorifique* to carry the world's first mechanically refrigerated cargo—chilled meat exported from Rouen, France, to Buenos Aires. He used a low boiling-point liquid for the refrigeration system, just as most refrigerators today rely on similar liquids to transfer heat away from a container storing food.

Tellier began experimenting with these refrigerants as a means of powering a

Illustration from Le Conquete Pacifique de l' Afrique par le Soleil, *by Charles Tellier. He predicted that his low-temperature solar pump would be useful for agriculture and industry, especially in remote, arid regions.*

solar pump. On the sloping porch roof of his shop in Auteuil, an exclusive suburb of Paris, he set up a row of ten solar collectors. They were metal plates, each four feet wide by eleven feet high, made of two sheets of corrugated iron riveted together. The grooves in the two sheets were aligned to form a series of hollow channels, through which the ammonia hydrate flowed. The bottom metal sheet of the collector was insulated to help block heat losses.

As the sun struck the top of the collectors, the metal conducted solar heat to the liquid inside. As a result, the ammonia vaporized and exerted a pressure of 40 pounds per square inch. The vapor circulated through pipes to a water pump consisting of a spherical chamber submerged in a well. The pressurized ammonia gas pushed a diaphragm in the chamber downwards, forcing water out of the pump in a jet. Afterwards the gas traveled through metal tubes set in a tank of cold water. The vapor condensed to a liquid again, and the ammonia was ready to repeat another cycle.

According to Tellier, the pump drew over 300 gallons of water an hour. But he realized that France's climate was "not favorable to the operation of such a device." He estimated that the capacity of the pump could more than double (to 792 gallons of water per hour) in the sunnier tropical regions of the world. Like his countryman Mouchot, Tellier looked south to the African colonies as the real proving ground for solar power. In 1890 he published a book entitled *Le Conquete Pacifique de l'Afrique par le Soleil* [The Peaceful Conquest of Africa with the Sun]. Its publication coincided with the partition of Africa by the European powers. Tellier understood that lack of cheap fuel hindered the underdeveloped continent's progress in industry and agriculture. He foresaw wide-reaching economic and social benefits from using low-temperature sun machines to make these lands more productive.

With this goal in mind. Tellier began testing designs for low-temperature solar engines. To increase the amount of solar heat collected, he put each metal plate inside a shallow wooden box covered with a single layer of glass. The hot box increased the collector's performance enough to power an engine, but more precise data on its operation are unavailable. A contemporary of Tellier, the noted engineer A.S.E. Ackermann, was one of those frustrated by the difficulty of ascertaining exactly how well the French inventor's solar devices worked. After reading Tellier's book, he commented: "With so much detail it is disappointing that [I] could not find the results of a single experiment with the [power] plant."

And when Tellier later dropped his promising solar research to return to the field of refrigeration, his reasons also remained something of a mystery. His experiments with low-temperature solar pumps had opened up a whole new approach to harnessing sun power; it remained for others to continue the work he had begun.

Willsie and Boyle

Two American engineers, H.E. Willsie and John Boyle, took up where Tellier had left off. Between 1892 and 1908 they explored the potential of low-temperature solar power plants based on the French inventor's design. In the May 13, 1909, issue of *Engineering News*, Willsie described what he and his colleague had accomplished. They had first begun to look into solar energy, he wrote, because they

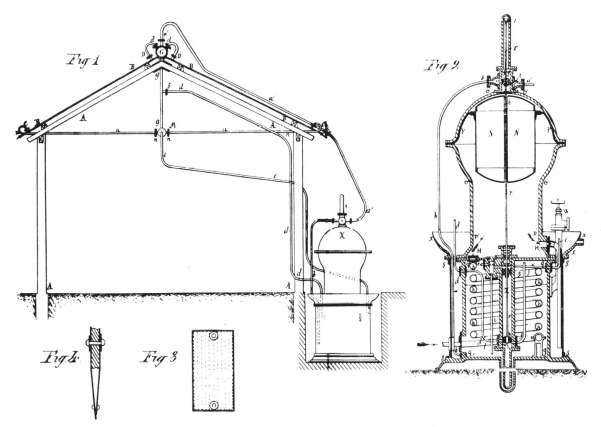

German patent drawings for Tellier's solar pump, 1885. In the cross-section of the water pump (right), pressurized ammonia gas pushed down on a diaphragm (NN); later the gas condensed to liquid ammonia in a series of coils (h) and returned to the solar collector.

realized—as had Ericsson and Eneas before them—that the sun-drenched American Southwest desperately needed a source of cheap power for its irrigation pumps and mines.

At the outset both men agreed to avoid publicizing their solar experiments until they had achieved some tangible results. They did not want to jeopardize their careers, for "the building of sun motors has not been an especially good recommendation for engineers." Frauds, failures, and eccentrics had given sun power a bad name. For instance, one priest proclaimed that his solar machine proved the truth of Genesis and that its use would serve as the cornerstone of a new social order!

Willsie and Boyle began their research with a review of the solar motors built by their predecessors. They discovered that "the state of the art most developed with reflectors to concentrate the sun's rays upon some sort of boiler." But they also knew that every reflector-powered motor had been a commercial failure. Therefore they chose to work with a Tellier-type motor using a low boiling-point liquid, so that a sophisticated and expensive reflector would be unnecessary.

The partners spent a decade drawing up blueprints and conducting small exper-

Cut-away view of the prototype hot-box collector built by Willsie and Boyle in Olney, Illinois. Solar energy warmed the water in this box enough to vaporize the sulphur dioxide that powered a low-temperature engine.

iments. In 1902, they decided the time had come to implement their plan for a full-scale sun power plant. Willsie explained the impetus for this decision:

> Boyle was then in Arizona, surrounded by conditions which daily remind one of the desirability of converting the over-abundant solar heat into much needed . . . power. To the Southwesterner "cheap power" brings visions of green growing things about his home to stop the burn of the desert wind, and of the working of the idle mine up the mountain side.

Before moving to the Southwest they built a crude prototype of their solar power plant in Olney, Illinois, to test an important modification of Tellier's original design.

In Tellier's process the solar collector plate had to be strong enough to contain the high-pressure vapor it produced. Willsie and Boyle felt that this would probably make the collector too expensive. Instead they decided to use water to transfer solar heat from the collector to a low boiling-point liquid in a separate system of pipes. Solar-heated water did not require such a heavy-duty collector, and the high-pressure vapor could be kept in a more limited circuit.

This innovation allowed them to eliminate Tellier's metal plate-hot box combination and substitute a simple hot-box collector—a shallow, rectangular wooden box covered by two panes of glass. The exterior walls and bottom of the box were insulated with hay; the hot-box interior was covered with black tarpaper. Three inches of water filled the box. Even in the cold October weather of the American Midwest, the solar-heated water was hot enough to vaporize sulphur dioxide, which then drove a low-temperature engine. This test proved that a sun machine did not need a solar reflector—surprising even Willsie and Boyle's friends who, Willsie noted, "were skeptical about window glass being able to take the place of mirrors."

In December of that year, they repeated the experiment in Hardyville, Arizona, where "over 50 percent of the heat reaching the glass was absorbed by the water." After a few additional tests, Willsie and Boyle patented the system in 1903. They also incorporated the Willsie Sun Power Company that year, which enabled them to procure capital by selling stock and obtaining loans.

The First Solar Power Plants

St. Louis became the site of a full-sized solar power plant built by the Willsie Sun Power Company in the spring of 1904. A set of large, shallow, rectangular boxes covered with glass served as solar collectors. The bottoms of the collectors were inclined toward the south, and held a thin film of water. As the sun warmed the water it travelled to a boiler, where ammonia was heated to a high-pressure vapor that drove a six-horsepower engine. Through condensation the ammonia returned to its liquid state and flowed back to the boiler. The water circulated back to the collectors in a separate cycle.

The plant ran on sunless days and at night as well, when an auxiliary boiler powered by conventional fuel took over. Newspapers in St. Louis and New York announced the success of this 24-hour-a-day solar-powered generator. Pleased with the results, Willsie and Boyle next moved their operation to the Southwest. They bought some land in the Mojave desert just outside of Needles, California—one of the hottest places in the country, where the sun shines 85 percent of the time.

Over the following years they built and rebuilt the Needles plant several times, each time making design improvements. Lack of money brought their work to a temporary halt in 1906. Although the interruption was unwelcome, they did learn some valuable lessons about the effects of the harsh desert climate on their apparatus. After two years of exposure only two percent of the glass collector covers had broken, and a few of the panes had turned purple in the sun. Aside from some minor damage, the power plant had weathered well. In the spring of 1908 additional funds enabled them to continue, and by mid-year the fourth and final version of the plant was completed.

The last Needles plant utilized a dual water-heating process that generated higher water temperatures. Angled to the south, two groups of hot boxes had a solar collecting area of over 1,000 square feet. The first group of hot boxes, with a single glass cover, raised the water to 150°F. The second group, with two glass covers separated by an air space, increased the water temperature another 30°F.

The hot water went to a boiler, where coils of pipe containing liquid sulphur dioxide were heated as the water passed over them. Willsie and Boyle chose sulphur dioxide and not ammonia for this motor because a "sulphur dioxide engine had been carefully tried out in Germany," said Willsie, "and because it required less heavy pipe and fittings to withstand the pressure . . . obtained in the heater." But sulphur dioxide had its dangers, too. If the chemical ever mixed with the heated water, a powerful acid would be formed that could quickly corrode the metal parts of the machine. But this did not present any immediate difficulties, and in test runs the vaporized sulphur dioxide performed well—generating a maximum of 15 horsepower.

The Needles plant also boasted one feature that other inventors had long dreamed of—a solar energy storage system. Mouchot had tried decomposing water into hydrogen and oxygen. Others had suggested that excess solar energy collected during the day could be used to lift weights or water; the force of their descent at

Below: Flow diagram for the last Needles power plant, from an article in Engineering News. *Solar-heated water (single-headed arrows) from the collectors (H) accumulated in a storage tank (S.T.) before being fed to the vaporizer (V) and there surrendering its heat to the sulphur dioxide. From the vaporizer, sulphur dioxide gas drove the engine (E) and then condensed back into a liquid at the condenser (C) before returning for another cycle.*

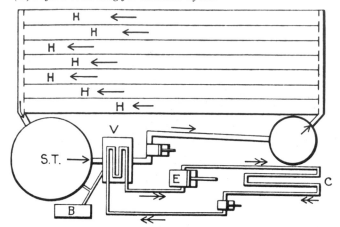

Willsie and Boyle's fourth and final sun power plant (above) built in Needles, California, in 1909. The hot-box collector sits in front, with the glass-covered boiler room behind it and the cooling tower to the left. Inside the boiler room (detail, top right), solar-heated water flowing over pipes vaporized the liquid sulphur dioxide inside them and drove a 15-horsepower engine.

night would generate power. Aware that increasing the temperature of one pound of water 1°F stores as much energy as raising a similar amount of water 778 feet, Willsie and Boyle chose to store solar-heated water as the simpler of the two methods.

The solar-heated water produced during daytime operation flowed from the collectors into an insulated tank. The amount of water needed at the time went on to the boiler; the rest was held in reserve. After dark, when a valve to the storage tank was opened, hot water flowed out and passed over the pipes containing sulphur dioxide, and the engine could continue working. Willsie could rightly claim, "This is the first sun power plant . . . ever operated at night with solar heat collected during the day."

The Needles plant worked better than any solar generator previously built. But one critical question remained: was it economically superior to conventional generators? Willsie noted that the sun plant would cost $164 per horsepower to build, versus only $40 to $90 for a conventional plant. But operating costs favored the solar plant, especially in the American Southwest where coal was very expensive. According to Willsie's calculations, running a conventional coal-fed steam engine cost $1.54 per kilowatt-hour, as opposed to only $0.45 per kilowatt-hour to operate the solar plant. As a result of the fuel savings, the solar plant could pay for itself in less than two years.

However, a couple of years after Willsie made this optimistic comparison the gas-producer engine was introduced to the Southwest. This type of engine burned coal to produce artificial gas, and motors using artificial gas were two to four times more efficient than conventional coal-burning engines. Thus, the solar engine became less economically attractive where coal could be cheaply shipped. Nevertheless, there were many areas where coal had to be transported from the railroad line by horse-drawn coach, raising the cost of operating a gas-producer engine. Willsie mentioned the example of a mine in Mojave County, California, where "the fuel is hauled 30 miles over the desert and then 5 miles over a mountain range," resulting in a coal bill of about $72,000 annually. In such remote areas the solar power plant would be highly marketable if mining companies and farmers could be convinced that the large investment was worthwhile in the long run. In the Mojave, Willsie noted, it took less than eight years to pay for itself through the fuel saved. He also pointed out that "during the summer months there will be an excess of sun power . . . that may be profitably used for the manufacture of artificial ice."

Despite the favorable prognosis, there is no record of the two inventors expanding their operations. And it is not known why their experiments ended with the Needles plant. Nevertheless, Willsie and Boyle made a giant stride toward the commercialization of sun power. They demonstrated that a solar reflector was not required to run an engine, and that a hot-box collector could easily drive a low-temperature motor. Their solar energy storage system combined with a conventional engine as a backup enabled a solar power plant to operate around the clock and throughout the year.

A parabolic-trough reflector and boiler at the sun power plant in Meadi, Egypt.

Chapter 9
The First Practical Solar Engine

While Willsie and Boyle were designing their Needles power plant, a self-taught engineer on America's East Coast was also exploring the use of hot-box collectors to drive low-temperature engines. A resident of the Philadelphia suburb of Tacony, Frank Shuman had an eye toward the technology of the future. When he took his children to that turn-of-the-century wonder, the picture-show, he told them, "Boys, the day will come when you will see this in your living room!" Not only did Shuman see television on the horizon, but like others before him he foresaw a time when fossil fuels would be extremely scarce and solar energy would become the industrial world's only hope. "One thing I feel sure of," he wrote, "and that is that the human race must finally utilize direct sun power or revert to barbarism."

Described by a professional journal of the time as a man "of large practical experience who [has] already made noteworthy and valuable inventions," Shuman entered the field of solar energy in 1906. After studying the sun machines invented by Mouchot, Ericsson, and Eneas, he reached the same conclusion as Willsie and Boyle concerning the reasons for their commercial failure:

> With the high temperatures involved, the losses by conduction and convection are
> so great that the power produced was of no commercial value. Where . . . mirrors
> are used, the primary cost . . . and the apparatus necessary to continuously present
> them to the sun, have rendered them impracticable.

Shuman therefore rejected the use of solar reflectors in the initial stages of his research and began experimenting with hot boxes. As de Saussure, Langley, and others had discovered before him, he found that hot boxes could reach temperatures high enough to boil water.

Next, Shuman began testing a low-temperature solar engine resembling the one used by Willsie and Boyle. He first built a one-foot-square hot box with blackened tubes inside that held ether, a low boiling-point liquid. The solar-heated ether vapor drove a tiny engine, the kind that were commonly sold at toy stores at the time for a dollar. Shuman tried using a similar collector to run an engine somewhat larger than the first, and was able to produce one-eighth horsepower.

Shuman saw he was on the right track, but it would take a lot of time and work to refine his ideas. As he put it, "Natural forces are not entirely conquered in a few years." It would also take a substantial amount of capital. To attract investors, Shuman realized that he would first have to build a successful demonstration engine—just as it had taken solid evidence of the possibility of flight to attract investors to the fledgling aviation industry:

> You will at once admit that any businessman approached several years ago with a
> view of purchasing stock in a flying machine company would have feared the sanity
> of the proposer. After it has been shown conclusively that it can be done, there is
> now no difficulty in securing all the money which is wanted, and very rapid progress
> in aviation is from now on insured. We will have to go through this same course.

With money he had made from other profitable business ventures, Shuman built a demonstration sun motor in his backyard. In August of 1907 he printed up handbills inviting the public to "attend an exhibition run of the first practical solar engine—on any clear afternoon between 12 and 3 p.m. during the next two weeks." Although Shuman, like Ericsson before him, was exaggerating his

SOLAR POWER   LIQUID AIR

MECHANICAL POWER, HEAT, LIGHT,
ELECTRICITY, REFRIGERATION
AND FERTILIZERS

FROM SUN HEAT AND AIR

YOU ARE INVITED TO ATTEND AN

EXHIBITION RUN

OF THE

FIRST PRACTICAL
SOLAR ENGINE

AT 3400 DISSTON STREET

TACONY, PHILADELPHIA, PA.

ANY CLEAR AFTERNOON BETWEEN TWELVE AND THREE P. M.
DURING THE NEXT TWO WEEKS

PLEASE ACKNOWLEDGE RECEIPT AND SAY WHEN
YOU WILL COME

FRANK SHUMAN

TACONY, PHILADELPHIA, AUG. 20TH, 1907

Left: Frank Shuman, inventor and solar visionary of the early twentieth century.

Right: Handbill heralding the demonstration of Shuman's first solar motor.

achievement somewhat—for Mouchot, of course, had built an equally "practical" solar engine at Tours in 1874—the demonstration motor was a success. It continued running well past its original two-week billing, "working steadily on sunny days during the summer of 1907 and 1908," reported *Engineering News*, "pumping thousands of barrels of water." The engine also produced power on sunny days during the cold Pennsylvania winters, with snow piled up around the collectors.

The power plant was a larger version of the low-temperature motors Shuman had tested earlier. The hot box, totaling 1,080 square feet of solar collection area, lay flat on the ground and contained blackened pipes in which a low-boiling-point liquid circulated. The solar-heated vapor operated an engine with a capacity of 3½ horsepower.

Combining Hot Boxes and Reflectors

Although the Tacony sun motor performed no better than Willsie and Boyle's plant at Needles, Shuman was much more persuasive than they in selling the idea of

Top: The first Tacony sun plant in operation, 1907. In this photograph from an article in Engineering News, *the hot-box solar collector is flat on the ground at left, while the engine and condensing coils can be seen at right.*

Left: Stock certificate for the Sun Power Company, founded by Frank Shuman and a group of investors to provide the capital for expanded operations.

solar power to wealthy investors. He painted a glowing picture of solar energy enabling industry and agriculture to expand in the fuel-short but sunny regions of the world—powering the nitrate industry in the deserts of Chile and the borax industry in California's Death Valley, and irrigating arid lands in the Australian interior, Eastern India and Ceylon, and the American Southwest. "Throughout most of the Tropical Regions sun power will prove very profitable," he predicted. "Ten percent of the earth's surface will eventually depend on sun power for all mechanical operations."

Shuman's confident, charismatic presentation of the case for solar energy convinced a number of American investors, who had made large profits on his other inventions, to gamble on his solar venture. They formed the Sun Power Company, and for the next several years Shuman was able to obtain enough financial backing to experiment with improvements in his solar system. His long-range goal was to build a large-scale solar power plant, and efficiency and economy were the crucial factors in making this ambitious project practical. He knew that the solar engine he had tested could be improved on both counts. As *Engineering News* observed, "Instead of developing a horsepower with only one or two square feet of heating surface area as in a locomotive boiler, or eight or ten square feet as in a stationary plant, a sun engine built [according to Shuman's first design] . . . would require one or two hundred square feet." Thus, a solar plant of industrial size would need so much collection surface in relation to the horsepower produced that the initial capital investment would be prohibitive. Shuman was determined to keep the purchase price of a solar plant low, for exorbitant first costs were to him "the rock on which, thus far, sun power propositions were wrecked."

Shuman developed several ways to increase the efficiency of his solar engine while holding costs down. But he needed an additional infusion of funds to move from the drawing board to actual construction of a large-scale solar plant. He found the necessary capital abroad—a group of British businessmen agreed to form the Sun Power Company (Eastern Hemisphere) in 1910. Shuman announced that "sufficient money is at hand to go into business on a large scale, and there will be great developments in the near future."

The first "great development" was the decision to develop and construct the largest solar plant ever built. The company chose Egypt (then a British protectorate) as the eventual site; land and labor were cheap there, and the desert sun was strong. However, the plant was to be first constructed in the United States to allow Shuman to test it thoroughly before shipment to Africa. He built the plant on two thirds of an acre near his home in Tacony, Pennsylvania. One of his primary objectives was to increase the amount of heat produced in the collectors. He accomplished this by adding reflectors to the hot boxes and installing a mechanism to adjust the angle of the collectors for optimum solar exposure. He also put metal heat absorbers similar to Tellier's inside the hot boxes and improved their insulation.

In all, there were 572 collectors with a total area of 10,276 square feet. The reflectors he used were plane mirrors made of ordinary silvered glass, and their construction required no expertise in optics. Two mirrors, each 3 feet high by 3 feet wide, concentrated sunlight onto each horizontal hot box of similar dimensions—for a reflector to collector area ratio of 2 to 1. These concentrating collectors sat adjacent to each other in 26 long rows, so that each row resembled a trough. In each row there were 22 collectors lined up horizontally to form the trough's base; the reflectors on each side formed the trough's walls.

To increase heat collection even further, the rows of collectors were tilted so that the glass covers of the hot boxes were always facing directly into the noonday sun. Notches in the frames on which the collectors were mounted enabled Shuman to adjust the angle of the hot boxes and mirrors every three weeks so that they stayed in alignment with the sun.

Besides increasing heat collection, Shuman redesigned the engine. He had concluded that his previous engine, which ran on the pressurized vapor produced by a low boiling-point fluid, did not generate enough power. But substituting high-temperature steam did not seem like a good idea either, because that would

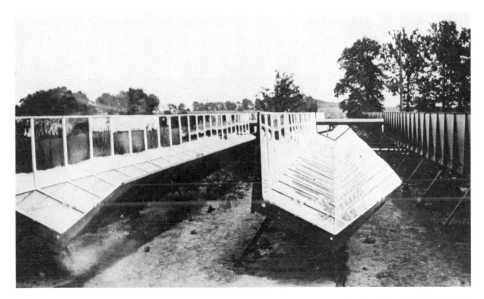

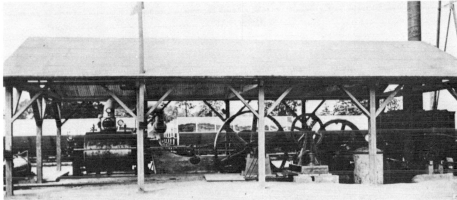

Top: Concentrating solar collectors in the second Tacony power plant, 1911. Mirrors on two sides reflected additional sunlight into each hot box.

Left: The engine in Shuman's second power plant was driven by low-pressure steam produced in the boiler at left.

require increasing the temperature in the collectors to the point where the overall efficiency of the system would decrease because of large heat losses.

Instead, Shuman invented a special motor that ran on low-temperature, low-pressure steam. Just as water boils at a lower temperature in the mountains than at sea level because there is less air pressure at higher altitudes, this engine was able to convert hot water into steam at a temperature below 212°F because the steam was kept isolated from atmospheric pressures. Each row of collectors had a feed pipe at one end through which cold water entered the channels of the metal absorbing plates in the collectors. The sun-heated water flowed out through a pipe at the other end of the row, and the hot water pipes from all the rows emptied into a main duct leading to the engine. After the water was converted to low-temperature steam, which then drove the motor, a condenser converted the steam back into water that returned to the collectors.

This engine generated more power than any other solar motor previously built. Connected to a pump, the device could deliver 3,000 gallons of water per minute—

The second Tacony power plant in action. This sun machine pumped over 3000 gallons of water per minute to a height of 33 feet.

raising it 33 feet. Almost 30 percent of the solar energy striking the collectors was converted into useful heat, producing a maximum of 32 horsepower and an average of 14 horsepower on a typical sunny day. The noted engineer A.S.E. Ackermann, sent by the British partners in the company to oversee the design of the plant, estimated that the engine's performance would improve by 25 percent in the hotter, more consistently sunny climate of Egypt.

Solar Energy in Egypt

Before Shuman shipped the solar plant to Africa, the British investors requested that Professor C.V. Boys, an eminent physicist, be brought in to review the project. After investigating the plant's operation, Boys pointed out that only the top of the hot boxes collected solar energy, while the bottom not only had no solar exposure but actually lost heat. He suggested that the reflector surround the hot box on three sides so that the bottom would receive solar rays as well. This could be accomplished by replacing each row of 22 concentrating collectors with a single parabolic,

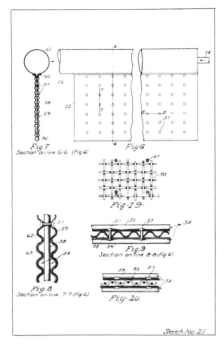

Professor C.V. Boys' modification of Shuman's second solar collector (below). Diagram (left) shows structural details of the boiler situated at the focus of this parabolic-trough reflector.

trough-shaped reflector, inside which a long, glass-covered boiler was suspended. As a result of this design change, the ratio of reflector area to hot box area would be increased to 4½ to 1, more than twice the concentration ratio of the previous plant. Shuman and his fellow engineers readily approved the change. At the time they thought Boys' design was original; later Ackermann discovered that it closely resembled Ericsson's parabolic-trough concentrator.

Rather than rebuild the entire system to incorporate this change and then transport it to Egypt as originally planned, the company decided to construct an entirely new plant on site. In 1912 Shuman and his crew arrived in Meadi, a small farming community on the Nile, 15 miles south of Cairo. Five solar collectors were built on a north-south axis, each one measuring 204 feet long and 13 feet wide with a distance of 25 feet between them. Shuman used construction materials that were readily obtainable in the industrialized world, yet sturdy enough to last many years. The collectors rested on a foundation of reinforced concrete and were supported by steel frames. Additional measures were taken to ensure that the collectors could withstand gale-force winds.

Each trough-shaped reflector was made from sections of silvered glass, held in place by small brass springs that absorbed the stress of expansion and contraction due to temperature changes. The boiler running down the middle of the trough was held up by rods capable of absorbing additional thermal stress. Covered by glass, the boiler was 15 inches deep with a flat bottom and trapezoidal sides. Inside, in a 3½-inch zinc pipe extending the length of the trough, water was heated to 200°F. This water was converted to low-pressure steam capable of powering an engine similar to the one used in Tacony.

The reflectors shifted automatically throughout the day so that the sun's rays

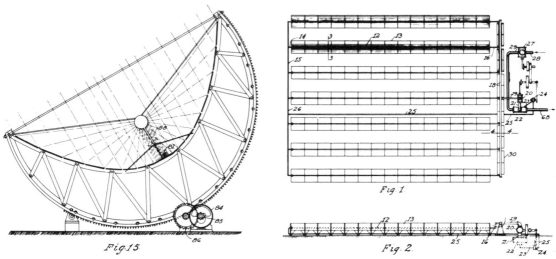

Fig.15   *Fig. 1*

Fig. 2

*The sun power plant at Meadi, Egypt (top). Plan of the solar collectors and
pumping equipment (above right). Patent drawing showing how a thermostat
(81) controlled motors that kept the reflector aligned with the sun (above left).*

were always focused on the boiler. When the reflector was facing directly into the
sun, a thermostat located directly behind the boiler remained in the shade. As the
sun moved westward and its rays struck the thermostat, a tracking mechanism
sprang into action and moved the reflector a fraction of an inch toward the west.

The Meadi plant could operate 24 hours a day. A large insulated tank, similar to
the one used by Willsie and Boyle, held excess hot water for use at night or during
overcast or rainy days. This enabled the engine to drive a conventional irrigation
pump at all hours and in all weather, further increasing the efficiency of the plant.

Shuman set up a public demonstration of his sun-driven engine in late 1912. But
the boiler reached temperatures too close to the melting point of the zinc pipes.

Consequently, the metal began to melt until, according to one observer, the pipes "finally hung down limply like wet rags." The trial run had to be postponed while the zinc pipes were replaced with cast iron.

By July 1913, the plant was again ready for testing. Such luminaries as Lord Horatio Herbert Kitchener, Consul-General of Egypt, were invited to witness Egypt's first display of solar power. Gentlemen in pith helmets and Panama hats and women carrying parasols to protect their fair skin from the tropical sun watched the giant solar machine swing into action. It produced more than 55 horsepower, enough to pump 6,000 gallons of water a minute. The absorbers collected 40 percent of the available solar energy, which was much better than the Tacony plant. This machine far surpassed the performance of previous solar engines, including Willsie and Boyle's plant in 1908 and Eneas' conical reflector engine in 1904.

Shuman's solar engine compared very favorably to a conventional coal-fed plant. True, the solar plant still had an enormous ratio of collecting surface to horsepower produced—exceeding 200 square feet per horsepower; and the purchase price was double that of a conventional plant—$8,200. Nevertheless, the costs of running the solar plant were so much less than the operating costs of a conventional plant that the financial outlook for solar power was extremely bright. With coal going for $15 to $40 a ton in Egypt, the solar plant would save over $2,000 annually in fuel costs. In two years the extra investment in the sun plant would be paid off; in four years the entire purchase price would be met; and in subsequent years the plant would operate for practically nothing, with the exception of expenses for repairs and labor.

Future Plans Fail

After devoting seven years of work to solar power and spending nearly a quarter of a million dollars, it seemed to Shuman that his earlier optimistic predictions were beginning to come true. He wrote in February of 1914:

> Sun power is now a fact and no longer in the "beautiful possibility" stage. . . . [It will have] a history something like aerial navigation. Up to twelve years ago it was a mere possibility and no practical man took it seriously. The Wrights made an "actual record" flight and thereafter developments were more rapid. We have made an "actual record" in sun power, and we also hope for quick developments.

Many others agreed, avidly supporting solar power. Some were former skeptics like those at *Scientific American*, who now praised Shuman's solar engine as "thoroughly practical in every way."

Besides the world of science, Western Europe's colonial powers also lauded Shuman's work and looked forward to the enormous economic benefits of using solar energy in underdeveloped Africa. Lord Kitchener offered the Sun Power Company a 30,000-acre cotton plantation in the British Sudan on which to test solar-powered irrigation. The German government called a special session of the Reichstag to hear Shuman speak, an honor never before bestowed on an inventor. Speaking in the German he had learned as a boy, Shuman described the fantastic possibilities of solar power and showed movies of the Meadi plant at work. Duly impressed, the Germans offered $200,000 in Deutschmarks for a sun plant in

Right: Egyptian peasants diverting the water pumped from the Meadi power plant.

Below: Officials and dignitaries attended the grand opening of the power plant in 1913.

German Southwest Africa. With such enthusiastic demonstrations of support, Shuman now expanded the scope of his plans. He hoped to build 20,250 square miles of reflectors in the Sahara, giving the world "in perpetuity the 270 million horsepower per year required to equal all the fuel mined in 1909."

But his grand dream disintegrated with the outbreak of World War I. The engineers running the Meadi plant left Africa to do war-related work in their respective homelands, as did Shuman who returned to the United States. He died before the war's end.

Gone was the driving force behind large-scale solar development. Moreover, with the Germans in defeat and their African colonies taken over by the Allies, the promises made to the Sun Power Company were as worthless as the Deutschmarks offered it. And the British, too, had lost interest in solar power. They began to turn

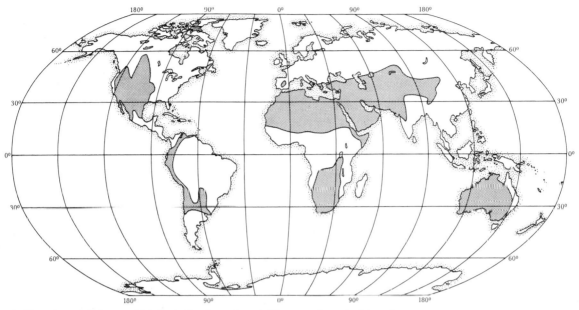

Arid regions of the world (shaded) where Frank Shuman thought his solar engine was a practical power source.

towards a new form of energy to replace coal—oil. By 1919, the British had poured more than $20 million into the Anglo-Persian Oil Company. Soon afterwards new oil and gas strikes occurred in many parts of the world—southern California, Iraq, Venezuela, and Iran. These were almost all sunny areas where coal was difficult to obtain—areas targeted by Shuman, as well as Mouchot and Ericsson before him, as prime locations for solar power plants. But with oil and gas selling at near giveaway prices, scientists, government officials, and businessmen became complacent again over the energy situation, and the prospects for sun power quickly declined.

III
Solar Water Heating

A Maryland gentleman of the 1890's enjoys a steaming hot bath provided by his Climax solar water heater.

Chapter 10

Early Solar Water Heaters

Although regular bathing had been commonplace in ancient Rome, the practice died out almost completely during the Middle Ages; not until the nineteenth century did it return to Europe and America. During the 1800's, the requirements of personal hygiene, advances in technology, and greater material well-being all combined to increase the demands for hot water. Pasteur's germ theory of disease underscored the need for frequent warm-water bathing. With the introduction of iron plumbing and cheap manufactured soap, such home hygiene became much easier than before. People also needed hot water for washing cotton clothes—which were rapidly replacing the woolens worn by everyone but the gentry.

Unfortunately, water heating remained a laborious and time-consuming task for the majority of Americans, who lived in small towns and rural areas without the benefits of gas or electricity. They had to rely on wood, gasoline or coal-burning stoves to heat their water. As one homesteader recalled,

> You took just one bath a week, a Saturday night deal, because it was such hard work to heat water on the stove. You put the water in pots, pails, anything which would hold water and you could lift. It took a while for those old stoves to get going because the heat first had to penetrate through the heavy metal.

Some people attached a four-gallon water tank to the side of their stove, eliminating the need to crowd the top burners with pots of water. The tank was made of cast iron, with a lid that lifted off to allow cold water to be poured in and then scooped out after being heated. Where there was enough water pressure, a more efficient method was devised that heated the water faster and did away with the burden of having to carry water from the pump or tap to the stove. Water circulated directly from the household pipes into metal coils looped through the firebox of the stove, and from there to a holding tank attached to the side of the stove. But even with this system the water took time to heat, and according to one old timer, did not stay hot for very long:

> Once you got the fire going really good, you'd have to wait about 15 or 20 minutes as the cold water heated up. The hot water would naturally rise up into the tank. And the holding tank was not insulated. That was a real problem because the water in the tank would be cold within an hour or so.

Wherever water was heated—whether on top, next to, or inside the stove—the job of starting the fire and keeping it hot was a chore. After the wood was chopped and brought in or heavy hods of coal lifted, the fuel had to be kindled and the fire periodically stoked. There were also the unpleasant side-effects of smoke, ashes, dirt, and in the case of coal, foul odors. In the winter, families endured such nuisances anyway as part of the price of using the stove to cook hot meals and help keep the house warm. But in the summer, as one resident exclaimed, "It was torture just to be in the house with the stove on!"

In large cities the situation was a little better. There were gas heaters, which ran on "artificial" or "manufactured" gas made by baking coal in an airless environment. Artificial gas had only one half the heating capacity of natural gas, was not as clean-burning, and left oily residues. The most common type of gas heater was the "side-arm," so named because it was attached to the side of an uninsulated hot water tank. The side-arm was not automatic. It had to be lit with a match. The water

A typical turn-of-the-century water heater. Water was warmed in metal coils inside a cookstove and stored in the holding tank beside it.

took a while to travel through the heating coils inside the side-arm and into the adjacent tank. And when the water got hot enough, "the tank would start jumpin' and you knew it was time to shut it off," said one plumber who installed them. If you forgot,

> You might get your hand scalded or get a face-full of steam if you opened the hot water faucet. There were times when they would split a tank. We had this one house where this woman started [the] side-arm up and went uptown and when she came back the back of the building was blowed off!

Besides being dangerous, these early gas heaters were too expensive for many families to use. The price of artificial gas was about $1.60 per thousand cubic feet around the turn of the century. Taking inflation into account, it cost more than ten

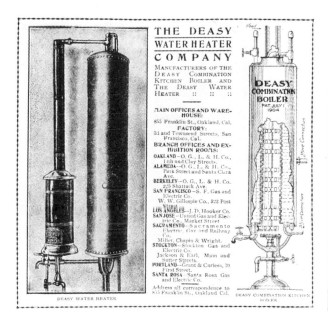

Left: A "side-arm" gas water heater, circa 1900.

Below: One of the first solar water heaters. These bare metal tanks were painted black and tilted facing the sun.

times what a family now pays (1980) for a quantity of natural gas with comparable heating capacity. As exorbitant as gas prices were, electric rates were even worse; nobody even considered heating water with electricity.

The Climax Solar Water Heater

Fortunately, a much safer, easier, and cheaper way to heat water was discovered —metal water tanks, painted black and simply placed where there was the most sun and the least shade. These were the first solar water heaters on record, and they worked. A prospector testified that sometimes "the water would get so damned hot you'd have to add cold water to take a bath."

The problem with these rudimentary solar heaters was not whether they could produce hot water but when and for how long. Even on clear, hot days it usually took from morning to early afternoon for the water to get hot. And as soon as the sun went down, the tanks rapidly lost heat because they were bare and unprotected from the night air.

These shortcomings came to the attention of Clarence M. Kemp, a Baltimore, Maryland, inventor and manufacturer. Kemp sold the latest in home heating equipment, including devices that produced artificial gas from coal for those living on large estates, and gas and coal stoves for the average homeowner. But fossil-fuel-consuming appliances weren't his only concern. In 1891 he patented a way to combine the old practice of exposing bare metal tanks to the sun with the scientific principle of the hot box, thereby increasing the tank's ability to collect and retain solar heat. Kemp called his invention the Climax, and it became the nation's first commercial solar water heater.

Kemp sold the Climax in eight sizes. The most popular model was the smallest, a 32-gallon heater that sold for $25 and measured 4½ feet long, 3 feet wide, and 1 foot

Right: Interior view of Kemp's factory in Baltimore, Maryland.

Below: Clarence M. Kemp, inventor of the Climax solar water heater, patented in 1891.

deep. The largest heater held 700 gallons of water and had a price tag of $380. Every model contained four long, cylindrical water tanks made of heavy galvanized iron painted a dull black. They lay horizontally next to each other inside a pine box insulated with felt paper and covered by a sheet of glass. The box was usually installed on a sloped roof or on brackets at an angle to a wall, so that the tanks lined up one above the other. The tanks were completely filled with water, which was then heated by the sun.

To draw hot water from the tanks, a faucet in the bathroom or kitchen was opened. In a house with pressurized plumbing, cold water from the inlet pushed solar-heated water out of the tanks and down to the bathtub or sink. If the home had gravity-feed plumbing, opening the faucet drew hot water from the tanks. Cold water refilled the tanks from a small reservoir located above the heater. A float valve in this reservoir allowed it to refill. In either system, a drain allowed the tanks to be emptied before the onset of freezing weather so that the water would not turn to ice and split the tanks.

Kemp advertised the Climax as "the acme of simplicity" compared with conventional heaters. Just turn on the faucet and "instantly comes the hot water," boasted the sales literature. Housewives could avoid the terrible heat of lighting the stove in the summer, and "gentlemen who occupy their residences alone during summer months, while their families are absent, can have the convenience of hot water without delay or attention." Of course, one of the main selling points was that a solar heater did not cost anything to operate.

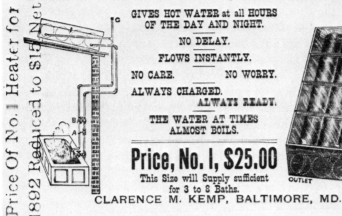

Climax Solar-Water Heater

UTILIZING ONE OF NATURE'S GENEROUS FORCES

THE SUN'S HEAT { Stored up in Hot Water for Baths, Domestic and other Purposes.

GIVES HOT WATER at all HOURS OF THE DAY AND NIGHT.

NO DELAY.

FLOWS INSTANTLY.

NO CARE.   NO WORRY.

ALWAYS CHARGED.
ALWAYS READY.

THE WATER AT TIMES ALMOST BOILS.

Price, No. I, $25.00

This Size will Supply sufficient for 3 to 8 Baths.

CLARENCE M. KEMP, BALTIMORE, MD.

OUTLET   INLET

Price Of No. I Heater for 1892 Reduced to $15 Net.

Left: Advertisement for the Climax solar water heater, 1892. The price of this, Kemp's smallest unit, had just dropped from $25 to $15.

Below: Two ways to install the Climax, from a company brochure: pressurized system (left), and gravity-fed system (right).

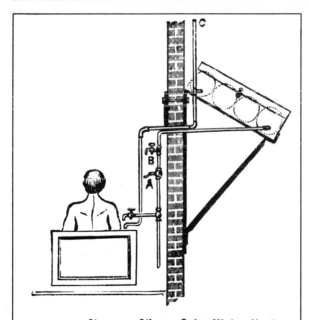

—Shows a Climax Solar-Water Heater supported by a bracket on the wall.

A.—Is the cock to use when the hot water is wanted. This passes cold water into the heater, displacing the hot water and forcing it through a pipe to the bath tub.

B.—Is the drain cock which is used to prevent freezing.

C.—The air opening which prevents vacuum in the heater and siphonic action.

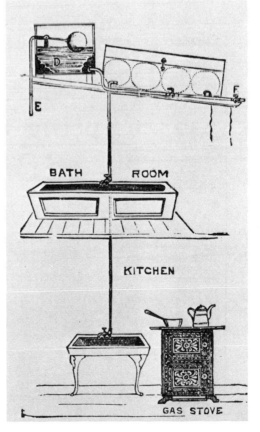

Solar Water Heating in California

In the Maryland area, Kemp claimed, the Climax could be used from the beginning of April until the end of October—producing water hotter than 100°F on sunny days even during early spring and late fall when daytime temperatures sometimes approached freezing. In areas of the country like California, the climate and fuel situation made the Climax even more attractive. Sunshine almost year-round meant free hot water most of the year, and extra savings because energy costs were high on the West Coast. California had to import coal at a price over twice the national average, and artificial gas was also expensive. As one journalist wrote, it was essential for Californians to "take the asset of sunshine into full partnership. A builder cannot afford to waste his sun rays."

Two Pasadena businessmen, E.F. Brooks and W.H. Congers, recognized the potential market for solar water heaters in southern California. In 1895 they paid Kemp $250 for the exclusive rights to manufacture and sell the Climax in California. Sales took off so quickly that just three years later, in 1898, Mrs. Sarah Robbins was willing to pay Brooks and Congers ten times what they had paid Kemp for just the southern California rights to the Climax. That same year Richard Stuart purchased the northern California rights for $10,000.

Climax installations spread from Pasadena to much of California and Arizona. By 1900 they topped the 1,600 mark in southern California alone. Economy was a

Various advertisements for Climax solar heaters that appeared in the Los Angeles Times *in 1900.*

prime lure of the Climax. For an investment of $25, the average homeowner saved about $9 a year on coal—and more if artificial gas was used for water heating. Landlords also considered the Climax a wise choice—like Samuel Stratton who outfitted his six flats with solar heaters. *The Pasadena Daily Evening Star* called Stratton "a level-headed businessman who knows a good thing when he sees it."

One satisfied Climax household, the van Rossems, had their solar heater on the southwestern side of the roof of their house (located near the present site of the Rose Bowl). Walter van Rossem, who was a child at the time, recalled that solar heaters became so popular that he and the others in the neighborhood did not think of them as anything out of the ordinary. "Everybody had one," he said. "There was nothing uncommon about it at all. I can't remember a house on the block that was built at the time or soon after that that didn't have a solar heater."

Van Rossem appreciated the Climax because he didn't have to fire up the stove very often to heat water. "What the heck," he confessed, "I didn't like to chop

Left: Four large Climax heaters supplied hot water to apartment dwellers in this building. The water tanks behind the collectors indicate that these were gravity-fed systems.

Below: The Pasadena home of Walter van Rossem, shown here sitting in his mother's lap. As early as 1896, this home had a pressurized Climax solar water heater—seen on the roof in this photograph.

wood any better than anybody else did!'' The rest of the family also appreciated the solar heater, though there were a few drawbacks. Van Rossem discussed how well the Climax performed:

> On an ordinary sunshiny day . . . by afternoon, my mother and our housekeeper would have enough hot water for baths and by evening there would be enough for us kids. Whether we had hot water the next morning depended on how much we used the night before. If we didn't use all the hot water up, it stayed fairly warm— enough to wash your hands and face.

As for laundry, van Rossem said the water was "hot enough for a small amount of washing, the things the women wore, but when we did the heavy washing, the stuff we kids wore like our overalls, we always had to boil water on the stove.'' Moreover, he noted, the seasons affected the amount of hot water available:

> In the wintertime usually there were a couple of kettles sitting on top of the wood stove heating. They were used for dishes and a lot of things because the water in the solar heater never got as hot in the wintertime as it did in the summertime.

Still, even on cloudy days "you'd be surprised how much it would heat up,'' van Rossem remarked.

The Walker Solar Heater

From the turn of the century until 1911, over a dozen inventors filed patents for improvements on the Climax. But only a few designs turned out to be technically and commercially successful. One of these was patented in the spring of 1898 by Los Angeles contractor and realtor Frank Walker. The Walker heater had only one or two cylindrical 30-gallon tanks. The tanks were set inside a glass-covered box, but the box did not protrude from the roof like the Climax—it fit inside the roof with the glass cover flush with the rooftop. This arrangement afforded somewhat better insulation and looked less obtrusive.

But the major advantage of the Walker heater over the Climax was that it was hooked into the conventional water heating system to ensure hot water at all times. At night or during inclement weather, cold water from the bottom of the solar water tank ran down a pipe to a heating coil inside the wood or coal stove or gas heater. Afterwards the heated water—which is less dense than cold water and rises naturally—flowed up through a second pipe leading to the top of the water tank. People found this method more convenient and cheaper because two sets of plumbing—one for the solar heater and one for the conventional heater—were no longer necessary.

The Walker cost less than $50, including installation. While it cost more than a similar-sized Climax, many customers throughout southern California willingly paid for the additional benefits.

The Improved Climax

In 1905 the rights to manufacture and sell the Climax in California were acquired by a branch of the Solar Motor Company—the firm founded by Aubrey Eneas.

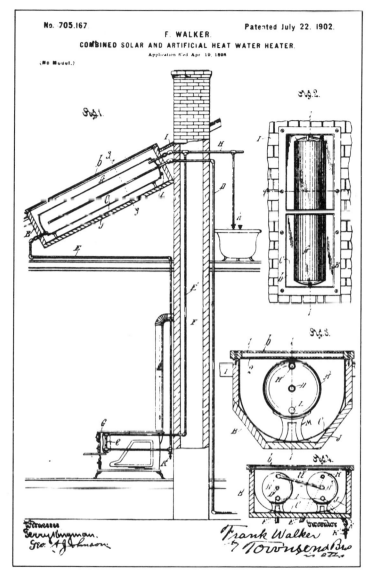

Left: Patent drawing for the Walker solar heater, 1902, showing how it could be linked to an auxiliary heat source—the kitchen stove.

Below: Solar inventor Frank Walker, who moved to Los Angeles from Canada in 1885. Well-regarded in the construction trades, he was a member of the city council by 1900.

Charles Haskell managed the Los Angeles headquarters of Eneas' business, which was listed under the name of the Solar Heater Company.

Haskell made a basic change in the design of the Climax water tanks. Noticing that it took many hours for the relatively deep body of water in the four cylindrical tanks to heat up, he decided to replace them with one large but shallow rectangular tank. It held the same total volume of water, but with less water per square foot the sun's heat penetrated more quickly and produced hot water earlier in the day. Like Walker's model, Haskell's was usually connected to a conventional water heating system that took over during unfavorable weather.

The Solar Heater Company called this updated model the Improved Climax. It was usually placed either on or in the roof, facing the direction with the best solar

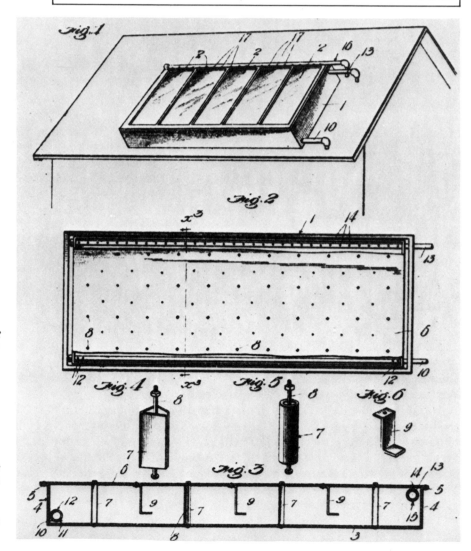

Top: Advertisement for the Improved Climax solar heater.

Right: Charles Haskell's patent drawings for the Improved Climax. This solar water heater had a single shallow metal tank that allowed the water inside to get hot much earlier in the day.

Six Improved Climax solar water heaters atop an office building in downtown
Los Angeles, 1907. Close-up (top) provides a better look.

exposure. According to one of the company's installers, the Improved Climax
worked quite well:

> Even on a foggy day, the first one to use it would get warm water. But of course, on
> a sunny day it would be much hotter. Why, hell's bells! You'd have to use the cold
> with it because you couldn't stay under the shower with just the hot water turned
> on. It really got hot!

Customers were just as laudatory. Los Angeles Superintendent of Buildings, J.J.

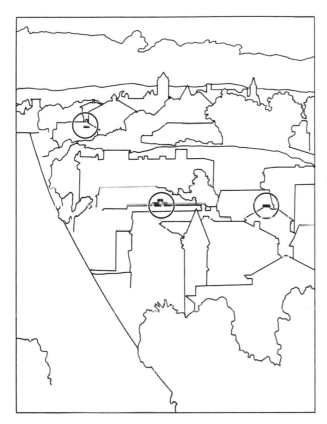

Olive Street in downtown Los Angeles, 1900 (opposite page). Note the Climax solar water heaters (diagram, left, and close-up, below) and the clear skies in the distance.

Backus, for example, wrote a testimonial that appeared in a 1907 issue of *The Architect and Engineer of California*:

> I take great pleasure in saying that after a thorough trial extending over a year and a half, our solar heater continues to give just as much satisfaction as when first installed. I am ready to admit that [at first] we were unreasonably prejudiced against the heater, and feel that refusing to let you install one in my house for so long a time after you first approached me upon the subject, we lost a great deal of comfort and convenience.

In southern California and in many areas further north, the Improved Climax and its predecessors, the Walker and the Climax, supplied large quantities of hot water for seven to eight months of the year—the Climax and Walker models heating water up to 120°F by late afternoon, and the Improved Climax reaching this temperature earlier in the day. But a serious defect hampered the effectiveness of these solar water heaters. While they lost heat less quickly than the early bare-tank heaters, their insulation consisted of only a pane of glass and a wooden box. The water did not remain hot for very long, especially on cloudy, cool days. Even under the best conditions, the water never stayed hot enough overnight to enable clothes to be washed in the morning. Kemp, Walker, and Haskell had brought the technology of solar water heaters a considerable distance in a decade and a half—but not far enough.

This Pomona Valley, California, family had a Day and Night solar water heater on the roof of their house in 1911.

Chapter 11
Hot Water— Day and Night

In the summer of 1909, in a little outdoor shop in the Los Angeles suburb of Monrovia, an engineer named William J. Bailey began selling a solar water heater that eventually revolutionized the industry. It supplied solar-heated water not only while the sun was shining but for hours after dark and the following morning as well—hence its name, the Day and Night.

Bailey had worked for Carnegie Steel in Pennsylvania before he moved west in 1908 to seek a cure for his tuberculosis. He soon discovered that his physician, Dr. Remington, had been experimenting with solar water heaters. To heat water faster and store the heat longer, Remington separated the solar heater into two units: a solar heat collector and a water storage tank. The collector consisted of coiled pipe placed inside a glass-covered box that hung on the south wall of his house. The small volume of water in the pipe heated quickly. And instead of remaining outside where it would readily cool down at night or during bad weather, the hot water flowed through pipes to a conventional water tank in the kitchen.

Bailey adopted Remington's idea of a separate collector and storage tank. But a good deal of heat still escaped from the tank because it was made of bare metal. For better heat retention, Bailey insulated the tank—a new concept in water storage for solar heaters, though by coincidence this was about the same time that Willsie and Boyle thought of incorporating an insulated water storage unit in their solar power plant at Needles. The average household-size tank made by Bailey held 60 gallons. It was encased in a wooden box, with powdered limestone between the box and tank at a thickness of 3½ to 9 inches along the sides and top. Bailey guaranteed that the water in this storage unit would not lose more than one degree Fahrenheit per hour.

The Day and Night collector was better designed than earlier heaters. A key feature was the use of copper pipes that held only a small amount of water, as in Remington's model. But even more importantly, Bailey added a metal absorber plate to transmit the solar heat accumulated in the hot box to the water in the narrow pipes.

The collector pipes and plate were enclosed in a glass-covered box measuring about 55 square feet. Only 4 inches deep, the box was lined with felt paper. Bailey put a large vertical cold water inlet pipe made of iron at one end of the box, and a parallel hot water outlet pipe at the other end. To these two headers he connected a series of smaller transverse pipes made of 5/8 in. copper tubing. The array of pipes was welded to a copper absorber plate at the bottom of the box, and both the plate and pipes were painted a dull black. Cold water entering the collector through the inlet pipe split into streams that flowed through the copper pipes. The streams heated up as the absorber plate conducted solar energy to the pipes. In southern California the system produced hotter water than previous heaters—100 to 120°F on sunny winter days, and 115 to 150°F during the other nine months of the year.

No pump was needed to circulate water between the collector and the storage tank. The Day and Night operated on the thermosyphon principle—that hot water is lighter than cold and rises naturally. The storage tank was located above the collector so that cold water in the bottom of the tank would be pulled by gravity down a pipe to the collector inlet. After it passed up through the copper pipes, the heated water rose through the outlet pipe and into the top of the water tank. This influx of hot water forced more cold water out the bottom of the tank and down to

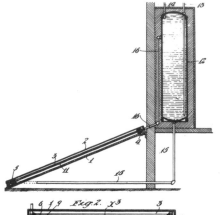

William J. Bailey, the engineer who developed and patented the Day and Night solar water heater (far right). Patent drawing of his first solar water heater, 1909 (right). This first "flat-plate" collector had a parallel grid of copper pipes welded to a copper absorber plate.

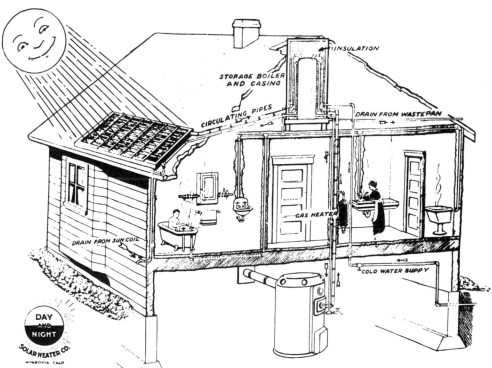

Cutaway drawing of a typical southern California home with a Day and Night installation. The storage tank sat above the collector so that the warmer, solar-heated water would collect there.

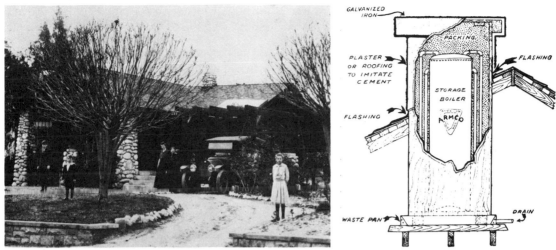

Left: Day and Night collector on the roof of a Monrovia, California, home—1915.

Right: The insulated storage tank used in the Day and Night system was usually disguised as a false chimney whenever it had to protrude through the roof.

the collector. The cyclic flow continued as long as the water in the collector remained hotter than the water in the bottom of the tank.

The collector was usually installed on the south-facing slope of the roof. It was a heavy piece of equipment and hoisting it into place was not easy. According to William Crandall, a Day and Night installer, the crew would tie a rope around the collector and "just by mean strength" pull it up on the roof on a set of skids. If putting the collector on the south side of a roof was impractical, they installed it as an awning, on the ground, or on brackets on the north side of the house in such a way that the collector would still get good solar exposure.

Installers usually placed the storage tank in the attic, making sure that the tank was higher than the collector. If the roof was too low, a hole cut in it allowed the tank to protrude above the roofline. The exposed section of the tank was then camouflaged with gray felt or roofing paper, holding-cloth, and stucco so that it resembled a chimney.

To ensure plenty of hot water during periods of bad weather or heavy use, Bailey advised customers to provide an auxiliary heater. The Day and Night could be connected to a wood stove, gas heater, or coal furnace. In places where people could obtain and afford electricity, a small electric insert heater was sometimes placed in the storage tank. This heater turned on automatically when the water dropped below the desired temperature.

Day and Night Captures the Market

Newspapers hailed the Day and Night as the "*ne plus ultra* of solar heaters," for it could "heat water and keep it hot under conditions that would render most other heaters of little or no use." Nevertheless, at $180, Bailey's invention faced stiff competition—a similar-sized Improved Climax cost much less. Before long, though, the Day and Night won over the buying public. Ned Arthur, one of the five original employees hired by Bailey in the early years of the business, elaborated:

> It was always a battle; if someone was going to buy a solar heater, Climax was on the job and I was on the job, and we fought like tigers. But after we got a few heaters in people saw the advantage of having water hot at night and Climax was out.

Unlike other solar water heaters, Bailey's product could supply hot water in the morning so that people did not have to wait until afternoon to do such chores as the laundry. "Many of our customers are reporting that they are putting out their entire washings early in the morning," one ad proclaimed. "Ask your neighbor if she can do this with her old-style heater." Furthermore, the Day and Night's long-lasting hot water enabled city residents to use their auxiliary gas heaters less frequently. Whereas the Climax reduced gas consumption by 40 percent, the Day and Night reduced it by 75 percent. In rural districts where there was no gas available, the demand for the Day and Night was also great because less coal or wood was needed.

The Day and Night soon began to edge out its rivals, but there was still a segment of the public that did not think any commercial solar heaters were worth buying. Paul Squibb, a rancher, spoke of the skepticism among some of his neighbors:

> When they first began putting in these Day and Night jobs, old timers were giggling about how silly they were. They'd say, "You couldn't get water any hotter than if you just stood a can full of water out in the sun." One poor guy nearly got the skin burned off himself. He said he'd put his hand in any water heated by the sun and the poor guy got an awful roasting. He jerked his hand out before he lost his skin!

Demonstrations set up at fairs and at the Day and Night office gave others a chance to test the company's claims of "steaming hot water day and night." One advertisement challenged,

> Step in some cloudy morning following a day of sunshine, hold your foot in the water from the heater for five minutes, and we will give you the heater. Cork legs are barred from the test.

Such publicity stunts helped the Day and Night to catch on, and in 1911—two years after Bailey had first opened up shop—the company was incorporated and moved to a larger plant.

The Sun Coil

Bailey's business continued to grow. But in the winter of 1913, a freak cold spell that hit southern California nearly pulled the company under. "Lowest temperatures ever known were reported in the orange districts early this morning. In some locations the thermometer registered 19 or 20 degrees with the mercury still falling," read the front page story in the *Los Angeles Times* on January 13th. The water inside many Day and Night collectors froze, and the copper pipes "popped like popcorn all over the county," said Day and Night employee Ned Arthur. Bailey's son, William J. Bailey, Jr., recalled that his father's telephone "rang all

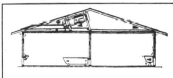

SOMETHING NEW

in line of

SOLAR HEATERS

Hot water all night

Hot water early mornings

Hot water 24 hrs. a day during sunny weather

Hot water on a rainy or totally cloudy day following day of sunshine

Day and Night Solar Heater Co.

205 S. Myrtle Avenue

(1 block north of the Park)

or

Any Reliable Plumber

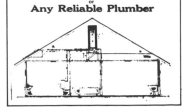

Top: Five rural homes in southern California with Day and Night systems (detailed above).

Left: Advertisement from the Monrovia News—*July 10, 1909.*

Top: Employees of the Day and Night Solar Water Heater Company pose outside the factory, circa 1912.

Above: An array of Sun Coil collectors set up at a California county fair.

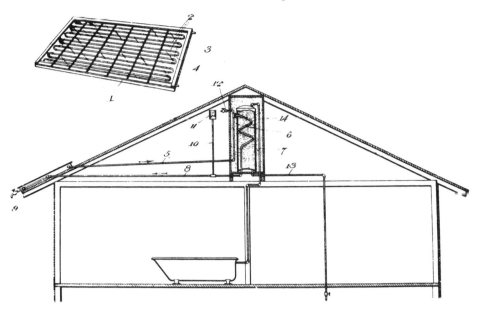

*Patent drawing—
the improved Day
and Night system.
An antifreeze so-
lution carried the
solar heat from
the collector to a
a coiled heat ex-
changer inside the
storage tank. "Sun
Coil" collector
detailed at top left.*

night long—irate customers were having problems with water coming through their ceilings. That sent him back to the drawing board."

Bailey came up with a way to prevent such disasters. Instead of water, a nonfreezing liquid—usually a mixture of water and alcohol—heated in the collector and traveled to a coil inside the storage tank. The heat passed from the coil to the water in the tank, and the cooled liquid returned to the collector for another round. This method resembled the process in Willsie and Boyle's engine: the liquid that flowed through the collector gave off its heat to another liquid outside the collector without the two ever coming into direct contact. Aside from taking care of the problem of freezing, making the collector's circulation system independent of the hot water supply in the storage tank had another advantage. In areas with very hard water, the collector could be filled with a mixture of alcohol and distilled water so that the tubing would not beome encrusted with mineral deposits.

Bailey made some other improvements in the Day and Night collector at this time. He did away with the iron headers and transverse pipes, substituting what he dubbed the Sun Coil design. A series of parallel pipes running horizontally were connected to each other at alternate ends with U-shaped fittings, so that they formed one continuous length of tubing in a zig-zag configuration. The tubing was then soldered to the copper absorber plate. This arrangement made the flow of water through the collector more efficient. The water entered at the bottom of the coil, and as it heated it rose up more quickly and out the top of the collector. Later Bailey switched from using copper pipes for the coil to ¾ in. galvanized steel, which was more widely available.

The Industry Expands

With a circulation system that eliminated the danger of frozen water pipes, Day and Night's business flourished once again in southern California. Sales began to

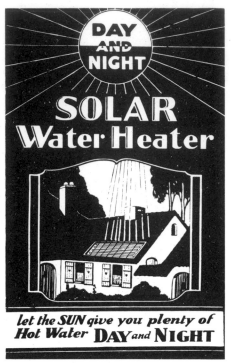

Left: Advertisement from the Arizona Magazine, *1913.*

Right: Day and Night sales brochure, circa 1923.

take off in other parts of the American Southwest, too. Two brothers, C.M. Eye and H.A. Eye, bought the rights to the Day and Night for Arizona and New Mexico in 1913 and set up their headquarters in Phoenix. They wasted no time in marketing the heater. By the following year a reporter for *Arizona Magazine* acknowledged that "the sight of the Sun Coil is becoming as familiar on Salt River Valley homes as in California, where they have been in general use for years." Soon Day and Night heaters also spread to Hawaii. But in northern California, solar water heaters were not as readily accepted. As Ned Arthur described the problem,

> It was awfully hard to break in. Everywhere I went . . . people would say, "Oh well, they'll work in southern California but they won't work up here."

Determined to prove the Day and Night's feasibility, Arthur arranged a demonstration in Palo Alto—near San Francisco. It was "right on the main street in the heart of town," he recounted. A sign attracted the attention of passersby with the message, "Hot Water From California Sunshine—Try It!" An arrow pointed to a faucet connected to a Day and Night heater. It was a convincing test—painfully so, for "people would come by and scald their hands. I sold heaters right and left," said Arthur. This marketing strategy worked equally well in dozens of rural towns in northern California.

As Day and Night's reputation grew, the manufacturers of the Walker and Climax heaters were forced out of business. The solar water heater became

SOLAR HEATER···
reduces Gas Bills

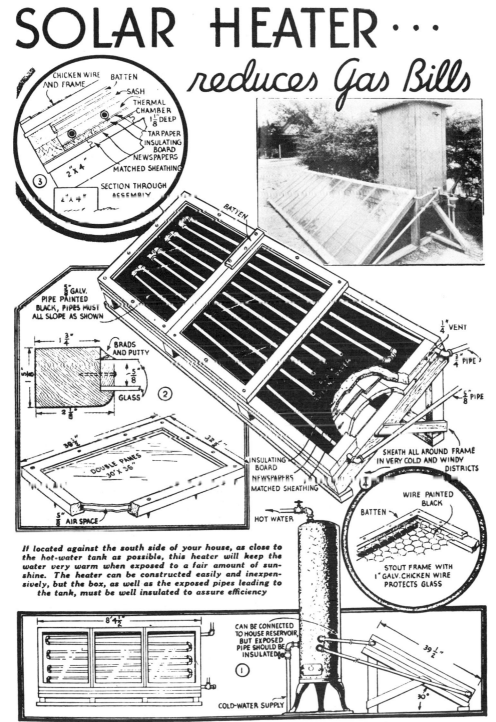

3 CHICKEN WIRE AND FRAME — BATTEN — SASH — THERMAL CHAMBER 1⅛″ DEEP — TAR PAPER — INSULATING BOARD — NEWSPAPERS — MATCHED SHEATHING — 2″x4″ — SECTION THROUGH ASSEMBLY — 2″x4″

BATTEN

⅝″ GALV. PIPE PAINTED BLACK, PIPES MUST ALL SLOPE AS SHOWN

2 1¾″ — BRADS AND PUTTY — ⅝″ — GLASS — 5/16″ — 2⅛″

¼″ VENT — ¾″ PIPE — ⅝″ PIPE

SHEATH ALL AROUND FRAME IN VERY COLD AND WINDY DISTRICTS

DOUBLE PANES 30″ X 36″

30½″ — 32¼″

INSULATING BOARD — NEWSPAPERS — MATCHED SHEATHING

⅝″ AIR SPACE

HOT WATER

4 WIRE PAINTED BLACK — BATTEN — STOUT FRAME WITH 1″ GALV. CHICKEN WIRE PROTECTS GLASS

If located against the south side of your house, as close to the hot-water tank as possible, this heater will keep the water very warm when exposed to a fair amount of sunshine. The heater can be constructed easily and inexpensively, but the box, as well as the exposed pipes leading to the tank, must be well insulated to assure efficiency

1 8′4½″ — CAN BE CONNECTED TO HOUSE RESERVOIR, BUT EXPOSED PIPE SHOULD BE INSULATED — 39½″ — 30° — COLD-WATER SUPPLY

Do-it-yourself plans for a solar water heater from a 1935 issue of Popular Mechanics. *Inset photo shows a unit built from these plans.*

Top: Sixteen Day and Night solar water heaters installed in a single subdivision of Monrovia, California. By 1918, there were over 4,000 in use.

Above: Home in Salt River Valley, Arizona, with a Day and Night installation.

synonymous with Day and Night's name. Bailey's only remaining competition came from local plumbers and do-it-yourselfers. Because the basic design of the heater was not very complicated, "the man who was handy with tools and pipe wrenches could build his own," as a trade magazine commented in 1914. One plumber in Ramona, California, sold collectors made of galvanized iron pipe coiled in a zig-zag pattern on an absorber of black tarpaper set inside a glass-covered box. A Santa Barbara plumber built a collector out of coiled pipe strapped to a copper absorber instead of welded to the plate as in the Day and Night model. People made storage tanks out of ordinary hot water tanks, which they boxed in and insulated with any coarse, dry material such as sawdust, ground cork, or rice hulls. Such heaters never had much impact on Day and Night's business, though, and by the end of World War I over 4,000 Day and Night heaters had been sold. In 1920 alone, over 1,000 people bought Bailey's system.

But 1920 turned out to be the peak year for sales of solar heaters. Between 1920 and 1930, huge discoveries of natural gas were made in the Los Angeles basin. Gas production soared and fuel prices plummeted. By 1927, consumers could get natural gas for about a fourth of what artificial gas cost in 1900. Networks of new pipelines brought cheap natural gas to urban and rural areas that formerly had no gas supplies.

Instead of trying to buck the trend toward gas, Bailey decided to capitalize on it. He began selling a Day and Night gas water heater that eliminated the objectionable features of the old-fashioned side-arm. His gas heater used some of the techniques that had made the solar heater such a success. A copper heating element conducted

Signal Hill oil field in the Los Angeles Basin, circa 1923. Plentiful oil and natural gas brought an end to the use of solar water heaters.

gas heat to the water, just as the copper absorber transmitted solar heat to the liquid in the solar collector pipes. And the gas heater's storage tank was now insulated, as was the solar tank. Moreover, Bailey added a thermostat that automatically heated the water to the desired temperature. "No trouble, no fuss—simply turn the dial indicator," ran a Day and Night ad promoting its new gas heater. "All the hot water you need, heated quickly and kept hot in an insulated tank, constantly awaiting your needs."

The gas companies provided economic incentives to customers buying gas water heaters—one of many programs they initiated to encourage gas consumption. "They'd finance gas water heaters on a monthly basis or let you carry 'em for a year or two," a retired plumber related. "The gas company would do anything to get you buyin' from them." In addition to easy terms, they offered cut-rate prices and free installation.

It was an unbeatable combination—cheap, accessible supplies of gas, the convenience of an automatic heater, and financial breaks that made purchasing a gas heater much easier on the pocketbook than paying for a solar heater. Solar water heaters were abandoned and new purchases of the Sun Coil dropped drastically. In 1926 Day and Night sold only 350 solar heaters; four years later, a meager 40 were installed. As Ned Arthur put it, ''Whenever a gas main would run out into the country, our solar heater sales quit.'' Bailey's old slogan of ''steaming hot water day and night'' now meant water heated by gas, and his company became one of the largest producers of gas water heaters in the nation.

Day and Night continued building and selling solar heaters in California—but at a greatly reduced level. Bailey's company sold a total of over 7,000 heaters before it stopped manufacturing them at the beginning of World War II. The last production run was made in 1941, according to William J. Bailey, Jr., and would have carried solar water heating technology more than halfway to Australia—but fate intervened:

Pan American Airlines bought a big lot of them and had intended to ship them out to the South Pacific to put them on Canton Island. That was the time when Pan American flew the old Clipper Ship runs to Australia, and Canton Island was the stopover point. They wanted hot water there and using solar was the only way they could get it. Those water heaters were on the dock in San Francisco, ready for shipment, when Pearl Harbor came along. They were never shipped.

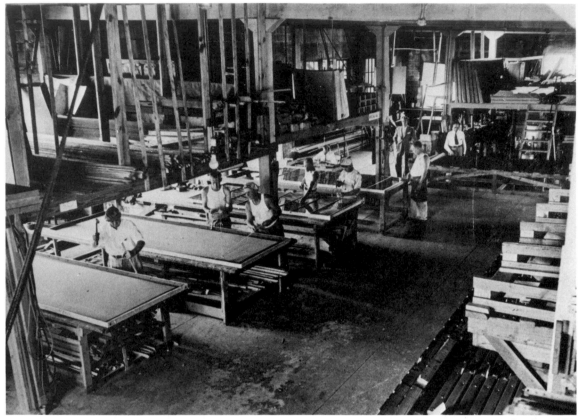

Workers assemble collectors at the Solar Water Heater Company in Miami, Florida, 1936.

Chapter 12
A Flourishing Solar Industry

The 1920's may have marked the beginning of the end of solar water heating in California, but a new market was then opening up in southern Florida. Post World War I Miami was a boomtown, with land speculators and realtors converting swamps into building sites for thousands of new hotels, apartments, and houses. The magnet of the sun and sea and the completion of a highway from New York to Florida brought vacationers and new residents to the area in droves. Miami's population almost tripled in five years—from 29,000 in 1920 to more than 75,000 in 1925.

But newcomers to this tropical paradise found that there was no cheap way to obtain hot water. California's huge gas discoveries did not ease the fuel shortage in Florida. Natural gas was unavailable there and artificial gas could not be supplied directly to homes and hotel rooms because it was difficult to lay pipelines through the subsurface marshland. Some people bought bottled artificial gas, which cost $4.60 for the equivalent heating capacity of a thousand cubic feet of natural gas— about a third more expensive than artificial gas in California. The more common alternative was to use electric water heaters. The big stumbling block was the high price of electricity—7¢ per kilowatt-hour, almost eight times its cost today, with inflation taken into account! Consequently, hot water was a luxury that many people did without.

H.M. "Bud" Carruthers, a wealthy builder who had come south from Pennsylvania, knew of the great demand for an inexpensive way to heat water in the Miami area. On a trip to California he hit upon the answer—the Day and Night solar water heater. Carruthers purchased the Florida patent rights from Bailey for $8,000 and an Oldsmobile touring car, and returned to Miami to set up the Solar Water Heater Company in 1923.

Carruthers began manufacturing and selling a solar water heater exactly like the Sun Coil in California: a collector made of steel pipe soldered to a copper plate inside a glass-covered box, with a separate insulated storage tank. And the response? As Harold Heath, who later became the company's sales manager, put it:

> Talk about a happy customer—they went nuts about this solar deal! People were living all this time taking cold baths or turning on electric heaters. The cost of operating those things would eat 'em up alive—they used 3,000 to 4,000 watts and the meter would just take off! In fact, some of our best customers were Florida Power and Light executives because they knew what it cost to heat water with electricity.

In contrast, a solar water heater saved so much money that it paid for itself in little more than two years.

People also liked the fact that solar heaters were automatic and appeared to be safe and reliable. In the words of one ad, "It can't stop working while the sun shines. It can't get out of fix, explode, or start a fire." By contrast, the electric heaters were dangerous. As Heath explained:

> If you wanted a bath you'd turn the switch to high and wait 15 or 20 minutes. Then you could take a bath. Nothing automatic about it. And if you got in your car and drove halfway to Palm Beach and somebody said, "Did you turn the hot water heater off?" and you said, "I don't know whether I did or not," you'd turn around and head back home, because the damn thing would blow up.

*Exterior view of the Solar
Water Heater Company in
Miami, 1924. The company
trucks are carrying large
bags of cork insulation.*

Consequently, "solar was real natural in Florida," said Heath. Carruthers did a
tremendous business installing heaters on new houses springing up along the coast,
and real estate agents and builders were eager to let prospective buyers know that
their homes were equipped with solar heaters. L.J. Ursem, a local realtor, ran the
following ad: "Don't fail to see this wonderful bungalow . . . garage with laundry
and wash trap, solar water heater, Frigidaire, other fine features." Carruthers'
company also put heaters on many hotels and apartment buildings, using a 14 ft
by 4 ft bank of collectors connected to a 300-gallon storage tank. Before long "there
seemed to be acres of solar collectors on apartment house roofs," commented
William D. Munroe, a resident of Miami at the time.

Because the Solar Water Heater Company owned the patent rights to the Sun
Coil in Florida, there was little competition. Precise sales figures are not known, but
Monroe observed that "the manufacture of solar water heaters was an established
industry in southern Florida by the mid-twenties." Carruthers built a factory
occupying an entire city block, and in April, 1925, the *Miami Herald* listed his
company as one of the seven largest construction firms in Miami.

But the solar heating industry was not destined to keep on climbing. In an
ominous foreshadowing of what would soon happen to the country's economy as a
whole, Miami's great building boom leveled off by early 1926 and completely
collapsed that summer. Then a ravaging hurricane swept the city in September.
"Things were all broke to hell," said Heath. Carruthers' business prospects were
bleak. They forced him to close up shop, and he went back to Pennsylvania—
entrusting his financial affairs in Miami to his associate, Charles F. Ewald.

The Duplex

Some five years later, around 1931, Ewald decided to try to crank up the solar
water heater business again. Since the Depression had kept the construction indus-

Top: Solar collector atop a typical Miami bungalow, circa 1924.

Above: Interior view of the new factory for the Solar Water Heater Company,
1925. Workmen are shown assembling solar collectors.

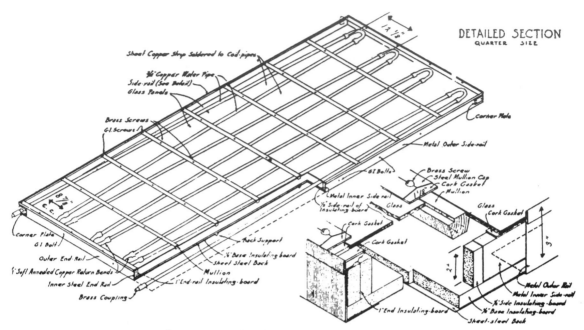

DETAILED SECTION
QUARTER SIZE

Construction details of Charles Ewald's all-metal solar collector, which used soft copper pipes to avoid ruptures from freezing.

try in a slump, he figured he would gear sales to owners of existing homes. But first he wanted to take a closer look at the Sun Coil to see if its performance could be improved.

Ewald began experimenting with new designs. One problem he tackled had to do with the effects of Miami's damp, humid weather. The wooden collector boxes and storage tank enclosures deteriorated quickly unless they were frequently repainted. To eliminate this expenditure of time and labor Ewald turned to all-metal construction, building the collector box out of galvanized sheet steel and the storage tank enclosure out of galvanized iron.

Better heat collection was another goal; he therefore insulated the sides and bottom of the collector box. He also replaced the steel tubing with soft copper—a better conductor of heat which was also more resistant to ruptures from freezing. Whereas Bailey had found that hard copper coils would split in very cold weather, Ewald reported that soft copper held up under test conditions as low as 10°F below freezing. The use of soft copper made it possible for heaters to be sold in places subject to occasional frosts without the special nonfreezing adaptation developed by Bailey. In 1958 an unusually severe cold wave in Miami verified the durability of soft copper—the coils suffered no damage, while nearly all the hard copper coils in other collectors burst.

Ewald also investigated ways to increase the efficiency of solar heat collection by rearranging the pipes inside the collector. After building several test models with different lengths of tubing and checking the water temperature in the storage tank, he concluded that one and a half feet of pipe should be used for every gallon of 140°F water desired. He also discovered that putting in a lot more pipe in the same zig-zag

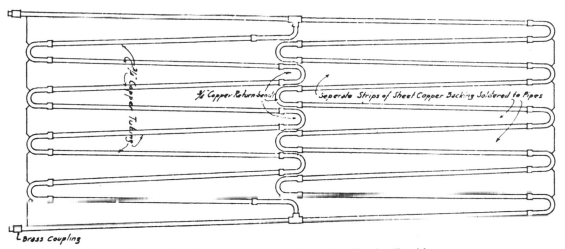

Piping pattern for the "Duplex" solar water heater patented by Charles Ewald in 1931. The two sets of coils allowed better water circulation.

configuration would not be practical for a household-size collector. A bigger collector would require a larger copper absorber plate and glass cover, and the added costs would have to be passed on to the customer. Even if people were willing to pay the price, the roofs of many homes were not large enough to accommodate a collector of great size.

Another possibility occurred to Ewald—spacing the tubing closer together so that it fit a collector of orthodox size. But this introduced a new problem—sluggish water circulation—which he realized when he watched how much time it took for pieces of cork floating in the water to circulate through the coil of tubing. The sharp-angled twists and turns of closely packed coils of pipe slowed down the water's rate of flow so much that only a small quantity of very hot water was produced. Ewald resolved the dilemma. He struck a balance between adding more tubing, spacing the tubing adequately for good water circulation, and keeping the collector box a manageable size. He put two coils of pipe instead of one in the collector, one on one side of the box and one on the other side. Each set of tubing was shorter than the single coil of the earlier model, but all together there was more pipe.

Ewald patented his new design as the Duplex. Because it produced hotter water in greater volume, the Duplex became the company's main product. However, the Sun Coil continued to be manufactured as well. But a Duplex collector, the company claimed, would provide 20 percent more hot water at temperatures 20 to 30°F higher than a similarly sized Sun Coil.

Another Ewald improvement was granulated cork used as insulation between the hot water tank and its metal shell. "Because it was a waste product there was no problem in getting it," according to Heath. "The cork was good insulation—it could retain the heat for 72 hours." As an auxiliary, the electric water heaters that residents already owned were piped to the solar storage tank.

When installing a heater the company found it important to determine the hot water needs of each household and to match the required tank size to the collector

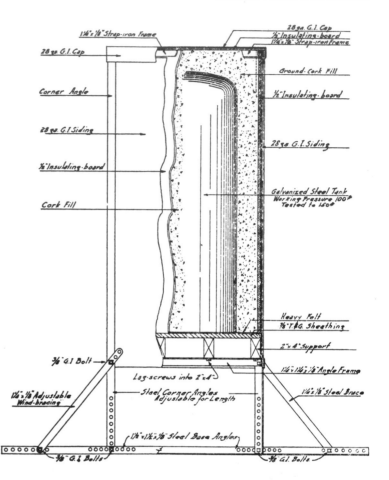

Right: The "new" all-metal storage tank used in the systems sold by Ewald's company.

Below: Harold Heath, sales manager for the Solar Water Heater Company.

size. Heath explained that a collector sized too small for the tank produced only lukewarm water; and if too large, the water would get excessively hot, which was uneconomical. He described how a typical retrofit job was assessed:

> The first thing we would do was case the house, find out how many people lived there, and then we'd button them down. "Now don't mislead us," we'd tell 'em, "We're selling you a hot water service, not just a water heater, so we have to know the amount of hot water your family needs. How many in your family? How often do you do laundry?" etc. Then we'd go from there to recommend the size of the system they needed.

Homeowners were satisfied with such tailored installation service and with the performance of the Duplex or Sun Coil. Between 1932 and 1934 a growing number of retrofit installations revived the solar heating business.

Federal Aid Boosts the Market

Suddenly, in 1935, the Solar Water Heater Company found itself amid an enormous wave of new construction in Miami. New Deal legislation passed by Congress

Typical Duplex installation, circa 1934. In this retrofit system the collector and storage tank were placed on the ground.

late in the previous year offered low-interest home mortgages through the Federal Housing Administration. A $3,500 home could be purchased for only $200 down and a small monthly payment. With a hungry market and federal funds picking up the tab, developers began building projects of 50 or 100 homes at a time.

However, the spate of new construction did not mean an automatic jump in solar heater sales. Most homes of the period were being designed with sloping roofs, instead of the flat roofs of the 1920's that had made installation of the storage tank easy—it was simply bolted to the roof. Now a builder had to construct a platform in the attic on which to set the tank, and cut a hole in the roof through which it could protrude. Moreover, a solar heater cost about twice the price of an electric heater. Nevertheless, consumer pressure soon persuaded developers to go with solar. A strong advertising campaign and word-of-mouth publicity made customers aware of the fact that with solar water heaters they would save a substantial amount on their utility bills. Economy-minded homebuyers started confronting developers over the issue, as Heath related:

> The buyers would ask, ''Where's the solar water heater?'' If the home didn't have one, they'd say, ''Oh well, we're not interested.'' It got to the point where they couldn't sell a house unless it had a solar heater.

Developers began to work together with the Solar Water Heater Company on installations in houses under construction. Usually the solar crew decided first

Top: A Florida bungalow of the late 1930's with a Duplex water heater installed in the roof of the car-port. The storage tank above the collector was disguised as a false chimney.

Right: The Solar Water Heater Company's Miami factory as it appeared in the middle of the 1930's. Ewald used the collec-tors and storage tanks on the roof for experiments that helped him improve system performance.

where the tank should be placed. The building crew then built the platform and casing for the tank, and plumbers ran the necessary pipes to the nearest location in the attic. The solar installers came in with the storage tank, setting it in its enclosure. Next they generally installed the collector on the roof, but sometimes put it up as an awning over a south-facing window—which served the dual purpose of providing hot water and helping to shade the house in summer. Then they connected the tank and collector to the plumbing lines, and the building crew finished off the job by disguising the tank as a chimney.

Federal programs stimulated not only the new-housing market for solar heaters but the retrofit market as well. With an FHA Home Improvement Loan a homeowner could buy a solar heater at 4 percent interest in installments of only $6 a month, with no money down. With monthly payments lower than normal utility bills for an electric water heater, people started saving money right after buying a solar unit.

The Solar Water Heater Company flourished, employing 34 workers to keep production rolling—three secretaries, five sales people, eight installing crews of two persons each, and ten in the plant to make coils, collector boxes and fittings. "We have no cause to complain about business," Ewald told the *Miami Herald* on August 5, 1935. "Only this week we have received from Pittsburgh our second carload of glass since March and another large shipment of copper tubing is on the way, which is the third we have ordered this year." Heath summed it up a little more colloquially: "We were selling heaters like bananas."

Competitors Crowd the Solar Field

Until 1935 Ewald's company was the only major solar manufacturer in the Miami area. But as the demand for solar heaters began to multiply, so did the competition. Plumbers and roofers branched out into solar; companies in allied industries such as U.S. Foundry—which worked with copper—also entered the field. Other firms such as the Pan American Solar Heater Company and Beutel's Solar Heater Company concentrated almost entirely on the solar heater business. In all, there were about ten important solar companies active in Florida by the late 1930's.

The Duplex patent still belonged exclusively to Ewald's Solar Water Heater Company, so it could not be copied by the new competitors. But this was not the case with the original Sun Coil, whose patent had expired by this time. Nearly everyone produced a facsimile, and rival companies promoted their products by giving them catchy brand names and emphasizing minor improvements in design. The Pan American Solar Heater Company advertised that its Hot Spot collector had flattened tubing rather than round pipe. They claimed this prevented shadows from being cast on the absorber plate.

Inevitably, some companies sought to reduce costs by cutting corners. A few soldered the copper plate to the coil of pipe only in spots rather than along its entire length, which resulted in poor heat conduction. A more common tactic was to eliminate the metal supports underneath the collector boxes. They laid the collector directly on the roof. But since the box was not watertight, rain seeped inside and leaked out the bottom. With no air circulation to permit evaporation of the water nor space for it to run off, water accumulated, rotted the roof and caused the collector

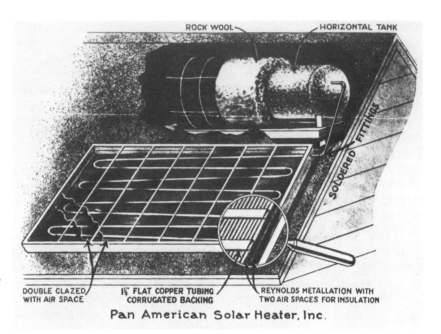

ROCK WOOL — HORIZONTAL TANK

SOLDERED FITTINGS

Details of the system marketed in Florida by Pan American Solar Heater. This company advertised that flattened copper tubing improved the performance of its "Hot Spot" solar collector.

DOUBLE GLAZED, WITH AIR SPACE   1½" FLAT COPPER TUBING CORRUGATED BACKING   REYNOLDS METALLATION WITH TWO AIR SPACES FOR INSULATION

Pan American Solar Heater, Inc.

box to rust. Other companies tried to capture the market by cutting down the size of the collectors. Unwary customers found out too late that these undersized collectors did not produce enough hot water.

People started to complain; they wrote to Washington since the FHA had financed most of the solar heaters. The government sent an investigator to Miami to check the veracity of their allegations. This induced the solar companies to form a voluntary association that adopted manufacturing standards—including proper guidelines for collector and tank sizes.

Such swift action from Washington and the solar industry restored public confidence in solar water heating, and sales climbed again. Estimates of the total number of installations made in the Miami area between 1935 and 1941 vary widely—from 25,000 to 60,000. More than half the Miami population used solar-heated water by 1941, and 80 percent of the new homes built in Miami between 1937 and 1941 were solar-equipped. More than 5,000 solar heaters were installed on large structures such as apartment houses, hotels, schools, hospitals, and factories, with more than half of them using 2,500-gallon storage tanks. The federal government purchased some of the largest solar heating systems, putting them in the officers' quarters at the giant naval air station in Opalaka, outside of Miami, as well as in the Edison and Dixie Court housing projects, which had a combined population of 530. In 1941 solar water heaters outsold conventional units in Miami by two to one.

Solar water heating also reached beyond the boundaries of Florida. In the late 1930's it spread to the Virgin Islands, where it was used in hospitals, and to Puerto Rico, Cuba and Central America. To the north of Florida, public housing projects from Louisiana to Georgia began using solar water heaters. At one large project in Georgia, for example, solar heaters supplied 480 dwellings with 35,000 gallons of hot water per day. After the first year of operation, one housing official reported

¾ MILLION DOLLARS SAVED

It may not seem possible, but it is a fact, 15,000 satisfied owners of the ALL-METAL DUPLEX SOLAR WATER HEATER actually saved ¾ Million Dollars the past year in hot water bills. You too, can save all the money you are paying for hot water by installing the All-Metal Duplex Solar Water Heater. It lasts a lifetime with no operating cost. If you are short of cash, use our easy Budget Plan. No money down, low monthly payments. Call us for more information. P h o n e 2-6496.

SOLAR WATER HEATER CO.

325 N. W. 25TH STREET   PHONE 2-6496   MIAMI, FLA.

EAST COAST BRANCHES:

FT. LAUDERDALE, FT. PIERCE, ORLANDO, VERO BEACH, PALM BEACH

Left: Advertisement for the Duplex solar water heater, circa 1937.

Below: Workman installing solar water heaters in the roof of the laundry room in a Florida subdivision built during the late 1930's. Every home in this tract had its own solar water heater.

The solar water heater even traveled to Cuba. "Without electricity, without gas, without coal, without cost!" states this sales brochure, "Hot water at all hours and for all uses."

that "solar water heating has proven to be most satisfactory, and while the initial cost is slightly more than for other types of hot water heating, the operation and maintenance costs are insignificant."

The Postwar Solar Decline

World War II halted the burgeoning solar heater industry. There was a government freeze on the nonmilitary use of copper, one of the main components of solar water-heating systems. But many of Miami's companies returned to business after the war—unlike California, where the already dwindling industry virtually disappeared. Despite the comeback, however, Florida's solar business never regained its earlier momentum. Hot water consumption began to leapfrog with the postwar baby boom and the advent of a newly affluent society that could afford dishwashers, washing machines, and similar home appliances. Families found that solar collectors installed on the basis of pre-war calculations of hot water usage were now too small. In an attempt to meet this increased demand, several solar companies introduced a thermostatically controlled electrical resistance heater as an auxiliary.

Insufficient hot water was not the only reason for a growing dissatisfaction with solar. Water tanks began to burst during the late 1940's and early 1950's, about ten years after they had been installed. According to Heath, "the tank would generally go around two o'clock in the morning when the city water pressure was up." Residents would awaken to rooms beneath the tank flooded with water. This badly tarnished solar energy's reputation of being a trouble-free way to heat water. The whole catastrophe could have been avoided if there had been better communication between the solar industry and the engineering community. As early as 1935 an article in an engineering journal explained that, to prevent corrosion, all parts of the solar system should be made out of the same kind of metal. Unfortunately most of the tanks were made of iron, and the pipes in the collectors were made of copper. The high temperature water promoted an electrochemical reaction between the two metals, which gradually ate away the tank until it ruptured.

The sharp rise in the price of solar heaters also alienated consumers. From 1938 to 1948 the price of copper doubled, and by 1958 it had tripled. Labor costs also skyrocketed. An unskilled solar-industry worker who had earned 25–40¢ per hour in 1938 made about $1.10 by 1955. The localized nature of the industry prevented large investments in labor-saving machinery, and it also took many man-hours to do the on-site installation work. Such high labor and material costs made the price of solar water heating systems shoot up from $125 in the 1930's to $350 in 1948 and $550 in 1958.

Meanwhile, electric water heaters became an economical and convenient alternative. Electric rates fell dramatically after the war, hitting 4¢ a kilowatt hour in 1948 and 3¢ in 1955. Many servicemen based in Florida during the war stayed on and the subsequent growth in Miami's population allowed the power company to charge less per customer, because its high capital investment in equipment could now be spread over a larger number of users. Florida Power and Light also pursued an aggressive campaign to increase electrical consumption, structuring rates to promote greater usage and offering free installation of electric water heaters. Demonstration heaters were displayed at its offices, and builders installing these and other electric appliances received promotional assistance.

The low initial cost of electric heaters also attracted customers. Large-scale production techniques had kept the price in line—in the early 1950's an electric water heater cost only $40 more than in 1938. Because the payback time on a solar heater system had ballooned to eight years by 1955, many found it cheaper to buy an electric water heater instead. And those who had the unfortunate experience of a ruptured tank were reluctant to pay more money for a new tank that might need to be replaced in another ten years. They preferred to change over to electric heaters, especially since the models now being sold were automatic.

With solar energy no longer the bargain it had once been, and electric water heating looking better and better, few people bought solar water heaters after the late 1950's. The industry became a service business—flushing out coils, fixing broken glass collector covers, and putting in new tanks for those who preferred to keep their solar water heaters because they were happy with what they were getting—free hot water.

IV
Solar House Heating

One of the apartment buildings in Siemenstadt, a solar-oriented workers' community near Berlin.

Chapter 13
Solar Communities of Europe

For almost a thousand years after the Fall of Rome, European architects virtually ignored the principles of solar orientation. The Classical writings on solar architecture of Socrates, Aristotle, Vitruvius and others fell into disuse. Moreover, large panes of glass went out of fashion; they were no longer as available, affordable or practical. A few examples of indigenous solar architecture continued to flourish in some areas of Europe, but the main urban centers were not planned for the sun.

In striking contrast, China's ancient heritage of urban planning and solar building design continued unabated. Indeed, solar city planning reached greater heights in China than it had in ancient Greece. Gabriel Magalhaes, a seventeenth-century Portuguese priest and one of the first Europeans to visit China, came back from the Orient filled with praise for Peking and other cities that were not built in what he called the "haphazard" manner of European towns. Homes were designed according to the cosmology that pervaded all aspects of Chinese culture. The south was associated with summer and warmth, the north with winter and cold. South was therefore the direction of health, and the preferred orientation of buildings was facing southward. Magalhaes reported that "all the cities and all the palaces of the king, the great lords, the mandarins, and wealthy persons are so built that the gates and principal rooms look toward the south." In the seventh century A.D. China brought solar planning to Japan—just as the Greeks had carried the idea to the Italian peninsula ten centuries earlier. Japan continued the practice even after the Chinese left the islands several centuries later.

In Mediterranean Europe and Asia Minor, folk architecture continued to apply some of the principles of solar building as a matter of common sense. Especially in Greece and Turkey, the idea of orienting homes to the south was not forgotten. As Theodore Wiegand, excavator of Priene, remarked, "The facades face south in villages and farms, as well as in the cities. Where they cannot face their houses south, the inner rooms are turned south." The Spanish also took the climate into consideration and built their homes with thick walls of limestone or adobe. This masonry helped to keep the buildings warm in winter and cool in summer. In the northern part of Spain the north wall of a house was always shut off from the cold winds.

An infatuation with Classical civilization, originating in Italy during the Renaissance, led to a revival of Greek and Roman architectural styles. Italian designers consulted the essays of Vitruvius and other ancient architects. They also worked for similar clients as had their predecessors in ancient Rome—people of wealth who could afford stately mansions and villas. The sixteenth-century Italian architect Andrea Palladio followed the advice of Vitruvius: summer rooms were built to the north side and winter rooms faced south.

When this Classical influence reached northern Europe, architects readily copied many of the outward forms of ancient buildings but neglected the solar principles that made the special beauty of these structures functional. They failed to orient buildings properly, missing an opportunity to heat them with the sun. Humphrey Repton, one of the few English architects to recognize the irony of this misuse of Classical solar architecture, remarked:

I have frequently smiled at the incongruity of Grecian architecture applied to buildings in this country whenever I have passed the beautiful Corinthian portico to the

north of the mansion house . . . such a portico towards the north is a striking
instance of the false application of a beautiful model.

Thus northern Europe's wealthy classes often had to heat their cold mansions
artificially while their prized peaches basked in the solar heat inside glass
greenhouses.

Solar Working-Class Communities

Of course, the lot of the urban poor during the bitterly cold winters of this era was
far worse. As peasants gradually migrated from the countryside to the cities, they
were often packed into congested, sunless rooms they could ill afford to heat. The
Industrial Revolution threw this migration into full speed. The first European nation
to industrialize, England was also the first to develop squalid slums that housed
armies of peasants forced to seek employment in the factories. In the mid-1800's,
Charles Dickens wrote about such workers' housing, built back-to-back, row upon
row, with little fresh air or sunlight penetrating the tiny apartments. Meanwhile, the
middle and upper classes could enjoy a winter's cup of tea in the sunny warmth of
the glassed gardens or conservatories just then coming into vogue.

With open sewers and a lack of hot running water, these crowded working-class
neighborhoods soon bred a host of lethal diseases. The dark, dank environment also
made it difficult to control epidemics of cholera, tuberculosis, smallpox and typhoid
fever. Sickness ravaged these slums in towns throughout England—taking
thousands of lives.

Physicians attributed this pestilence, in part, to the lack of direct sunlight. As the
ancient Greek medical authority Oribasius had advised centuries before, they held
that sunlight is essential to health. The motto of these environment-minded physicians was, "Where the sun doesn't go, the doctor does." Scientific discoveries
during the nineteenth century confirmed their belief. The French chemist Louis
Pasteur developed the germ theory of disease, and the prominent British physician
Sir Arthur Davies proved that ultraviolet light destroys bacteria. As epidemics
threatened entire cities—even the wealthier districts—liberal reformers and planners pressed the government to abandon its *laissez-faire* attitude toward the housing crisis. They demanded that something be done about the sunless, unsanitary and
overcrowded working-class ghettos.

England's predicament was repeated elsewhere in Europe. Industrialization
created intolerable living conditions for the new urban working class, and socially
concerned groups agitated for government intervention. By about 1900, many
countries had enacted public health and town planning laws ensuring sun-rights for
all citizens. Legislation passed by the British Parliament, for example, stipulated
that "rooms shall be placed as to permit sunlight to enter freely." A similar
provision in the German national building code stated that structures "must be
designed so that an adequate supply of sunshine can be secured in occupied
rooms."

As early as the 1860's, working-class communities were established in Great
Britain to provide healthy residences for industrial laborers. These communities
offered housing with plenty of sunlight, space, and adequate sanitation. Port Sun-

Left: The entire city of Ch'ang-an, capital of China during the T'ang dynasty, was built on a rectangular grid aligned with the cardinal points of the compass.

Below: Typical working-class slums of nineteenth-century London. Built back to back, row upon row, these dwellings had little access to winter sunlight.

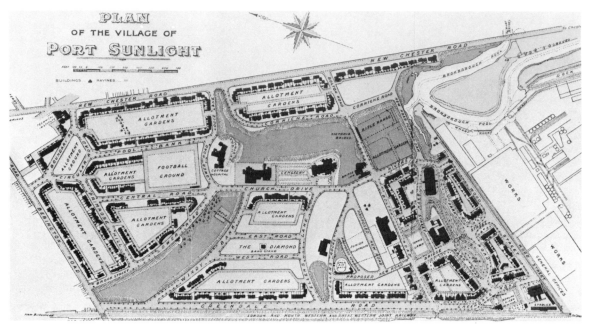

*Street plan of the working-class community of Port Sunlight, England. Wide
areas of open space ensured plentiful sunlight for every home.*

light, one of the earliest of these projects, was built by Lever Brothers for workers in
one of its soap factories. Compared with the sordid tenements of nearby Liverpool,
Port Sunlight was a godsend. And it was accurately named. As the developers
boasted, "The worship of sunshine is characteristic of every building in the village." The rows of houses were surrounded by wide areas of open space. According
to one observer, "Roads varying in width between 40 and 120 feet separate the
various blocks, so that air and light can penetrate [the homes] freely on all sides."

As planned working-class communities became popular in England and other
European countries, architects and planners began to study the question of solar
orientation more scientifically. During the early part of the twentieth century,
Raymond Unwin, an English urban planner, compiled information on the sun's
annual movements. He reached the same conclusion as the Greeks had centuries
before: "In taking the whole year round there can be no doubt that an aspect south
or slightly west of south may be considered the most desirable for dwelling rooms."
Whenever possible, he followed these guidelines in the design of homes for working-class communities. But the major purpose of Unwin and others of his time was
to obtain maximum sunlight for sanitation—not to collect solar heat. Ironically,
Unwin and his contemporaries did not know that window glass keeps out most of
the germ-killing ultraviolet rays. Still, solar heat during winter was a welcome
addition to homes built with a southern exposure.

The workers' communities did not bring about an immediate revival of the highly
structured solar communities common in ancient Greece, however. Reacting to the
horrors of congested urban slums that had usually been laid out row after row, most
planners sought to avoid such grid patterns. Without copying the Olynthian model

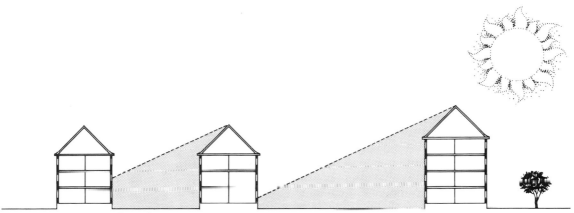

According to the studies of Augustin Rey, long apartment buildings facing south should be spaced apart 2½ times their height to avoid shadowing.

of arranging streets strictly on an east-west axis, it was difficult to build totally solar-oriented communities.

Urban Solar Planning in France

Most of the spacious, sunlit working-class communities that appeared in the latter part of the nineteenth century and the early twentieth century were built in the suburbs or the countryside. The possibility of extending the same benefits to urban residents inspired Augustin Rey, a French housing official, to investigate the feasibility of comprehensive urban solar planning. He began studying the effects of the sun's heat on buildings. From temperature measurements made throughout the year on surfaces of various orientation, he concluded that the most winter heat was available on southerly exposures. He also studied the amount of sunlight that would penetrate rooms under various conditions.

In 1912 Rey set out to discover the minimum amount of open space needed around a building to make sure that winter sunlight would not be blocked by an adjacent structure. He determined the sun's position on the winter solstice (December 21) for ten major cities of the world—Vienna, Paris, London, Berlin, Moscow, Washington, Philadelphia, New York, Chicago and Boston. Then he calculated the length of shadows cast by two, four and six story buildings. Rey found that at the latitude of Paris, south-facing apartments built one in back of another must have a space between them at least 2½ times their height. But buildings oriented north-south and facing east or west need to be spaced apart only 1½ times their height to avoid shading problems. Therefore, buildings facing south required nearly twice the land area of east or west-facing buildings.

Rey realized that these results raised serious questions about the possibility of solar access in urban areas. Paris, he concluded, already had a population nearly three times the ideal for proper solar planning. He felt that real estate speculation had already driven land prices in major cities to "criminal levels," and that only a government land policy opposed to "excessive speculation" would make solar architecture economically feasible in cities.

Tony Garnier's rendering of his Cité Industrielle, *an imaginary city of 35,000 inhabitants planned with equal solar access for all residences and other buildings.*

Rey discovered another obstacle to his plan for optimum solar access. A coherent strategy of planned urban expansion was necessary to achieve this goal. But in practically every European city, the idea of regulating urban growth was almost unheard of. To him, the need for rational city planning seemed "so evident that it would be infantile to ponder over it if one did not have to note the systematic neglect of this essential condition in the housing of urban residents." Only if city governments passed laws encouraging more rational city planning along with strong measures to limit land speculation, Rey argued, could solar communities have a fair chance in metropolitan areas.

Another Frenchman, Tony Garnier, envisioned building an ideal city of 35,000 from scratch—a city where residential neighborhoods and industry coexisted in health and harmony. Heavily influenced by studies of ancient Greek city planning, the young designer's "Cité Industrielle" bore a close resemblance to Olynthus and Priene. Garnier drew up extensive plans for a metropolis composed of long rectangular blocks with all buildings facing south. A house would occupy only half of its lot to prevent the shading of adjacent buildings during winter, and the remaining land could be used for private gardens.

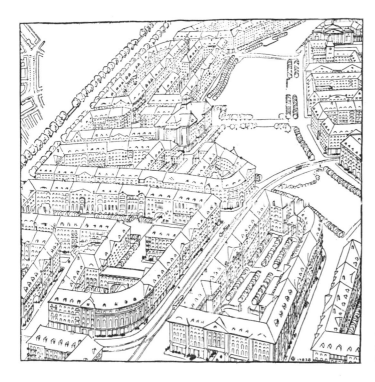

Prior to World War I, most apartments in German cities were crowded together in tight blocks.

Garnier and Rey's concern reflected a growing desire throughout Europe to find a humane alternative to the brutal living conditions that accompanied industrialization. Such idealists looked forward to a world in which human welfare took precedence over business and profit. While their plans never materialized, it was not long before similar dreams of other visionaries began to be realized.

German Solar Architecture

Following World War I, a new architecture emerged in Germany. A housing movement began to appear in which window glass was intentionally used as a solar heat trap to help keep buildings warm in winter. Postwar Germany was fertile ground for innovative trends. Defeat and humiliation at Versailles marked the death of the old Prussian culture. A few mourned its passing, but most intellectuals and artists of the new Weimar Republic celebrated the downfall of the archaic constraints of the past. They seized the opportunity for experimentation, forging new paths in many fields.

In architecture, too, the way was open for new concepts to take root. Designers chose architectural styles that were primarily functional—rather than merely aesthetic. With this philosophy, they attempted to build structures that would make practical use of the sun's heat. Germany had to face the harsh economic realities of a nation whose resources were being expropriated by the victorious Allies. In addition to requiring massive reparations payments that drastically inflated the Deutsche mark, the Allies occupied the Ruhr district—the heart of Germany's industry and source of most of her coal. According to the noted architect Marcel

With the Weimar Republic in postwar Germany came the Zeilenbau *plan, with its long, narrow apartment buildings oriented on a north-south axis.*

Breuer, a pioneer in solar design during this period, saving fuel with the sun was a major goal of the German housing movement. Wilhelm von Moltke, then an architectural student and now Professor Emeritus of Urban Planning at Harvard, agreed, pointing out that "sun orientation was considerably important because there is not much sunshine in Germany during winter, and, therefore, it was very much a concern to capture it."

The political environment during the 1920's also helped the development of solar housing. Left-leaning Social Democrats took power, and many German city officials sought to thwart land speculation by purchasing large tracts that were then selling at depressed postwar prices. These public properties became the sites of vast housing projects for the war-torn nation. This building explosion was aided by the efforts of cooperatives—groups of people who pooled their resources to buy land and build low-cost, high-quality housing. Sympathetic local governments often aided these housing cooperatives. Whether the land was owned by the government

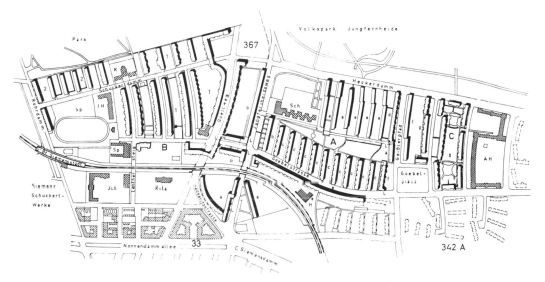

*Plan of the apartment complex of Siemenstadt, built in 1929 near Berlin. Most
apartment buildings in this workers' community faced east and west.*

or by cooperatives, architects were often able to implement their newfound solar
ideas.

The *Zeilenbau* Plan

Because of the acute housing shortage in Germany, large apartment complexes
were favored over individual homes. Traditionally, throughout most of Europe,
large apartments lined the perimeter of a block—facing in all four directions.
However, long narrow buildings several stories high were now being built in parallel
rows. Walter Gropius, a leading exponent of this architecture, explained that
"parallel rows of apartments have a great advantage over the old blocks, in that all
apartments can have an equally favorable orientation." But a problem arose here.
In which direction should these buildings face for best access to the winter sun?
When it came to determining the orientation of a building, "most architects made
their own rules," commented Marcel Breuer.

Most of the first solar buildings that went up in postwar Germany faced east and
west. As Rey had calculated earlier, this was the best arrangement with regard to
land economy. Called the *Zeilenbau* ("row-house") plan, this east-west orientation
became the pattern for numerous long, narrow apartment buildings that soon dotted
the German landscape. In 1929, for example, the city of Berlin helped finance
construction of the community of Siemenstadt, which eventually housed many
workers at the huge Siemens-Haskell Electric Works. The rows of four-story
buildings were erected far enough apart so that no apartment blocked another's
sunlight. The majority of rows ran north-south and were only two rooms deep. The
living room and a balcony usually faced west and a bedroom looked toward the east;

theoretically, half the main rooms received the morning sun and the other half got the evening sun.

The *Zeilenbau* plan, described as "heliotropic" by many, excited the international architectural community. Critics like Lewis Mumford reflected this enthusiasm:

> Above all, *Zeilenbau* permits the orientation of the whole community for a maximum amount of sunlight. In every other type of plan, a certain number of rooms will face north, but in *Zeilenbau*, when a correct orientation is established, every room and every apartment share equally in the advantage.

But despite this initially favorable reception, the *Zeilenbau* plan did not work as well as hoped. The winter sun is in the south all day—rising in the southeast, moving due south at noon, and setting in the southwest. Thus, the east and west windows received only modest amounts of sunlight on winter days because the sun's rays struck them at a glancing angle. And in summer just the opposite occurred—the bright rays of the morning and afternoon sun came straight into the east and west-facing rooms. This situation would have been even worse had Germany been located further south. Fortunately, many architects had included natural shading to shield the west windows from the summer afternoon sun. "It was a common practice to control the sun in the hot summer with trees on the west side," von Moltke recalled. During winter these deciduous trees shed their leaves, allowing some winter sun to enter the apartments.

The flaws in the orientation of *Zeilenbau* buildings soon became apparent. Paul Schmitt, a scientist who studied these structures, discovered that the streets collected more solar heat than the apartments! Schmitt branded the *Zeilenbau* plan as "so unfavorable that it must be considered ineffective." As experience with these mammoth complexes grew, their orientation shifted to the south. Henry Wright, a distinguished urban planner of the day, explained this transition:

> Up to the spring of 1931, the technique of community planning had been evolving rapidly toward a stereotyped repetition of building units facing almost exclusively in the one favored direction—toward the southwest. The advantage of the southwest exposure for Germany lies in the fact that the sunlight is very precious, due to the extreme northern latitude . . . and to the fact that most of the winter warmth is produced by the sun after the noon hour.

But here another problem arose. South-facing apartments had to be extremely shallow to take full advantage of the winter sun. This building constraint usually increased construction costs.

Back to a Smaller Scale

By the early 1930's many German architects began to feel that large-scale apartment complexes were an inherently uneconomical form of housing. The upkeep of such buildings offset their low initial costs. In addition, single-family homes—unlike apartment buildings—could be prefabricated, and mass production techniques could bring down building costs. With municipal governments and cooperatives providing cheap land, homes could be spaced further apart without

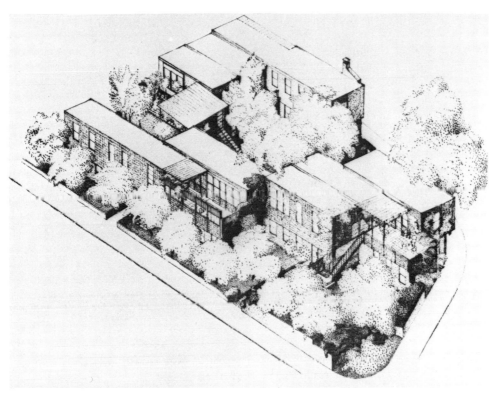

An apartment complex designed by Hugo Haring in 1934. As architects realized
the drawbacks of the Zeilenbau plan, they began to orient buildings south.

incurring excessive land costs. For these and other reasons, many architects began designing large communities of small buildings—rows of about six attached single-story homes.

Hugo Haring led the switch to single-story solar homes in the early 1930's. The rows of buildings ran east-west with the main rooms facing south, for as Haring put it, "There is no doubt any more that we, in the interest of obtaining proper solar radiation, must choose a south orientation." Large windows on the south side of the house admitted solar heat. There were few windows on the north wall since only cold winds, not sunlight, came from this direction in winter. In summer retractable awnings kept out the high summer sun.

The new movement had barely begun, however, when Germany was rocked by the economic and political upheavals that led to the collapse of the Weimar Republic in 1934 and the rise of Adolf Hitler. The Nazi Party lambasted the workers' communities, most of which were solar-oriented, as "communistic." In the name of returning to the honored German traditions, they ordered the construction of small, traditional country cottages for workers. According to Catherine Bauer, an architectural critic at the time, what really motivated the Nazis was their fear that building more of the workers' communities in the urban centers of Germany would make left-wing organizing easier. To dispel this threat, they employed a very "conscious and energetic program to turn potentially dangerous urban workers into

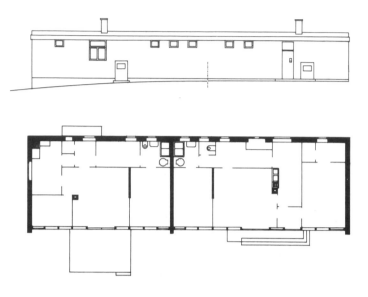

North side (top), floor plan (middle) and southwest view (bottom) of one of Hugo Haring's solar-oriented dwellings. There were only a few small windows on the north, east and west walls, but large expanses of glass on the south.

helpless pauper peasantry,'' according to Bauer. Thus the Nazis squelched any further expansion of the solar communities, and World War II soon destroyed much of what had already been accomplished.

Neubühl

After World War I, solar developments also sprang up in Holland, Sweden, Switzerland and elsewhere in Europe—although these movements never gathered as much momentum as the German movement. One of the largest and most sophisticated examples was the Swiss community of Neubühl—three miles south of Zurich.

Seven young architects organized Neubühl as a cooperative housing project. The 200 apartments ranged from small bachelor residences to family dwellings with six rooms. These units were apportioned among 33 separate structures perched along a mountain slope. The buildings faced south or slightly southeast, and were spaced far enough apart so that no building blocked another's solar access during winter. All apartments received the same exposure to the winter sun.

The living room and a majority of bedrooms in an apartment faced south. Glass spanned the entire south wall of each living room—and a good portion of the south

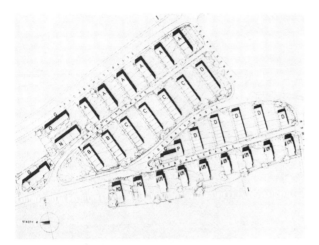

The Swiss solar community of Neubühl (below), near Zurich, as seen from the southwest. The plan of Neubühl (left) indicates that almost all apartment buildings faced slightly east of south.

bedroom wall as well. The kitchen, stairwell, bathroom and pantry occupied the north and northwest parts of the apartment. Hence, the kitchen remained cool in the summer—an advantage when the stove was in use. Retractable wooden shades and canvas awnings were rolled out to prevent the summer sunlight from entering the south-facing rooms.

In housing cooperatives such as Neubühl, personal welfare was the primary design motivation. But these communities were only the scattered indications of a much greater potential for solar development. After World War II, growing prosperity and cheap fossil fuels dampened any impetus toward further development.

The tiered rows of dwellings at Acoma allowed each home an unobstructed view of the winter sun.

Chapter 14
Solar Heating in Early America

American solar architecture began with an indigenous heritage. Some very sophisticated solar communities were built by the Pueblo Indian tribes of the American Southwest. During the eleventh and twelfth centuries A.D., the Anasazi Indians built a number of large community structures—some of them south-facing cliff dwellings and others out on open plateaus—that display a remarkable sensitivity to the sun's daily and seasonal movements. Such well-known ruins as Long House at Mesa Verde and Pueblo Bonito in northern New Mexico were built during this Classic period of Anasazi culture.

The ''sky city'' of Acoma is one of the most sophisticated examples of this solar architecture. Built atop a plateau as was the Greek city of Olynthus, Acoma has three long rows of dwelling units running east to west. Each dwelling unit has two or three tiers placed so as to allow every residence full exposure to the winter sun. Most doors and windows open to the south, and the walls are built of adobe. The sun strikes these heat-absorbing south walls much more directly in winter than in summer. By contrast, the horizontal roofs of each tier are built of straw and adobe layered over pine timbers and branches to insulate the interior rooms from the high, hot summer sun.

A study conducted by Professor Ralph Knowles of the University of Southern California proves just how well these Pueblo dwellings are suited to their climate. In winter, over one third of the sun's heat reaches the interior, while in summer only one quarter makes it inside the rooms. Even more surprising is the orderly town plan that guarantees all residents full, equal access to the sun's heat. Continuously inhabited longer than any other community in North America, Acoma has more than fulfilled the visions of its unknown preliterate designers.

The Spanish colonists who settled in the American Southwest often built according to a solar plan. Their dwellings took the form of single homes not very different from those common in many areas of Spain. The typical Spanish Colonial villa consisted of an adobe building oriented east-west with the main rooms facing south. The south wall absorbed sunlight during the day, and in the evening the absorbed heat was released from the adobe wall into the home. Shutters on the windows helped keep the heat inside the house at night. In summer, eaves sheltered the interior from the high, intense sun. Spanish Colonial architecture did not survive the later influx of Yankees from the eastern United States. With their ancestral roots in England and the rest of northern Europe, these immigrants did not understand how well-adapted the adobes were to their environment. They often replaced these structures with wood houses better suited to a New England climate.

But settlers in the eastern United States were not ignorant of climatic considerations in their own building designs. In the rural settlements of Colonial and Post-Revolutionary America, most dwellings fronted south—''as they ought to,'' said Thomas Breckenridge, author of *Modern Chivalry*, a picaresque novel of the time. In New England, the colonists often built ''saltbox'' houses facing toward the sun and away from the cold winds. These structures had two south-facing, windowed stories in front—where most of the rooms were placed—and only one story to the rear of the building. The long roof sloped steeply down from the high front to the lower back side, providing protection from the winter winds. Many of these saltbox houses also had a lattice overhang called a ''pergola'' protruding from the south facade above the doors and windows. Deciduous vines growing over the

Right: The plan of Acoma indicates that most of the dwellings were grouped in long rows facing south. One major exception is the Spanish Mission at bottom.

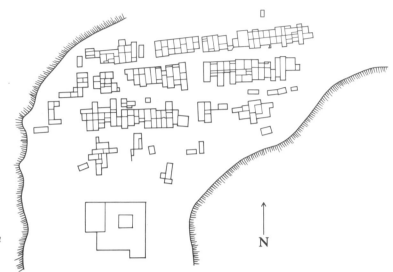

Below: Aerial view of Acoma Pueblo, New Mexico. Built in the twelfth century A.D., this American Indian village used intelligent town planning and building design to keep the homes warm in winter and cool in summer.

Left: The Colonial "saltbox" house was well-suited to the harsh New England climate. Deciduous vines on the trellis, or "pergola," blocked the summer sun but shed their leaves in winter to permit sunlight to enter the south windows easily

Below: An early California family in front of their Spanish Colonial adobe. The main, south-facing windows were shielded by overhangs from the hot summer sun.

pergola afforded summer shade but dropped their leaves in winter, allowing sunlight to pass through.

In the mid-to-late nineteenth century, solar building became a lost art in the United States. From the early 1860's to the late 1870's, nearly five million people poured onto America's shores—the vast majority of them ending up in cities like New York, Philadelphia, Boston and Baltimore. As the East Coast became increasingly urbanized, the common-sense designs developed by rural residents were all but forgotten. Where abundant open space had once allowed easy solar orientation, the arbitrarily planned street systems now made such orientation difficult. As

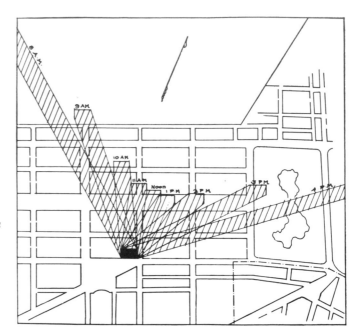

One of the studies by William Atkinson, showing the shadows cast in December by a 300-foot building. As a result of his work, the Boston city council passed an ordinance in 1904 limiting the height of buildings.

in Europe, cities on the eastern seaboard became crowded with multistory tenements inhabited mainly by poor immigrants.

Conditions in these cities were as bad as in the filthy, cramped, working-class quarters of European towns. In the 1870's, however, some of the more well-to-do began to desert the urban centers and settle in the surrounding countryside. But suburban housing was not much better. Architects like Bruce Price criticized the acres of "hideous structures" around New York and Philadelphia—suburban homes that were built shoddily, randomly oriented, and lacking adequate ventilation. They recalled a time when people built climate-sensitive homes such as the Colonial saltbox. In response to this trend, Price and a group of his colleagues began to design houses that remained naturally cool in summer and warm in winter. "The heat of the summer demands shady porches," Price declared, and "the cold of winter . . . sunny rooms." In his houses, he placed louvers so that summer breezes could cool the rooms; they also had low ceilings and well-insulated walls to minimize winter heat losses.

Urban Solar Access

The work begun by Pierce and others did not spread to the cities. As in Europe, land speculation put a premium on urban space, and the resulting high-density developments made solar exposure difficult. The housing crunch continued to tighten as the population hurtled upward. Some nine million immigrants landed between 1881 and 1900. With more and more people clamoring for housing in major cities, architects began building vertically.

Around the turn of the century the problem of solar access came to the attention of William Atkinson, a reform-minded Boston architect. The city's population had

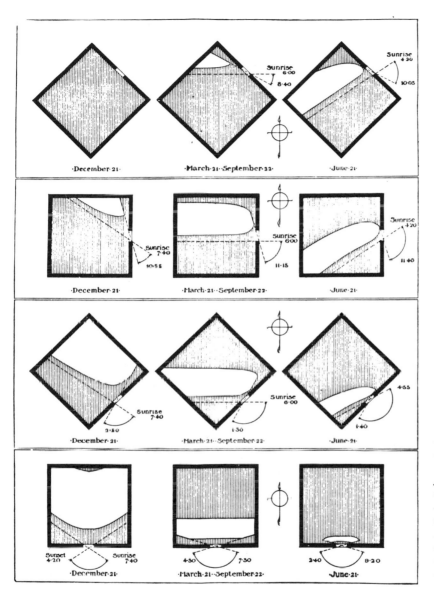

A chart Atkinson developed to show the solar penetration through windows of different orientation on the solstices (December 21 and June 21) and equinoxes (March 21 and September 22). From these studies, he concluded that a southern orientation was the best choice.

more than tripled in the previous 50 years, bringing a move toward taller and taller buildings. Atkinson saw that "the skyscraper enjoys an advantage of light . . . at the expense of lower and more ancient buildings." He drew diagrams showing how high rises shaded the surrounding buildings and presented his findings to the Boston city council in 1904. He convinced them that it was essential to guarantee access to the sun, and legislation was soon passed restricting the height of new buildings in Boston.

Atkinson had first become curious about ways to maximize solar exposure in 1894, when he was working on the design of hospital buildings. He found no rational criteria for the proper orientation of hospitals. So he decided to investigate how much sunlight penetrates windows facing in different directions. At the Harvard Observatory he found a table showing that the summer sun rises in the northeast, ascends high in the sky by noon, and sets in the northwest; in winter, the sun rises in the southeast, travels in a low arc across the southern sky, and sets in the southwest. East and west windows were the least desirable, then, since they received lots of sun in summer and little direct solar heat in winter. On the other hand, he realized, the opposite would be true of south windows.

In 1910, Atkinson conducted an experiment to verify his theory. He used a device he called a ''sun box'' similar to de Saussure's hot box—except that this box was mounted vertically, with the glass cover acting as a window and the box accumulating heat as would a room. The sun box had an inner wooden shell measuring one foot square that was covered by a single sheet of glass on one side. The other sides were surrounded by insulation, an air space, and an outer box. Atkinson built two sun boxes and exposed them to the direct rays of the sun. Between the months of June and December, he changed their positions periodically and collected data for orientations toward the west, southwest, south, southeast, and east.

The summer tests confirmed his hypothesis that east and west windows admitted too much summer sunlight. On June 21, for example, an east-facing box reached more than 120°F by 8 a.m.—52°F hotter than a similar south-facing box and 42°F

Cross-section of Atkinson's ''sun box.'' An inner box (E) with glass on one side was covered with insulation (D) and an outer box (B) on the other five sides.

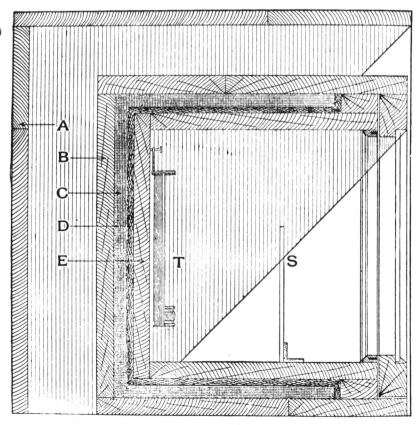

hotter than the temperature outside. By afternoon, a box looking toward the west registered over 144°F—44°F more than the south-facing box and 66°F over the air temperature. The most spectacular results, however, occurred on December 22. The temperature of the south-facing box rose to 114°F when it was only 24°F outside! The temperature in the boxes facing east and west never got much higher than the outdoor air temperature. Atkinson had proved that the most sunlight came from the south during winter and discovered that "the sun's rays are not of indifferent value in the heating of our houses."

In 1912, Atkinson built what he called a "sun house" for Samuel Cabot—a wealthy Boston aristocrat—to determine more exactly the heating potential of the sun. Actually a small shed, the building stood next to Cabot's spacious mansion. It was long and shallow, with glass spanning the entire south face so that the sun's rays could fully penetrate the interior. The other three walls and the roof had generous amounts of insulation to help retain the trapped solar heat. Atkinson reported that despite freezing weather outdoors, "a temperature of 100°F and over has been frequently attained within this building . . . entirely from the warmth of the sun's rays." The dramatic success of this sun house led him to exclaim, "Every dwelling may be converted into a sun box!"

Optimistic about the potential of solar heat, Atkinson published a book that same year entitled *The Orientation of Buildings, Or Planning for Sunlight*. But history did not affirm his optimism. Few American architects took advantage of his ideas about using solar orientation to obtain free winter heat. Atkinson's results were soon forgotten.

The "sun house" Atkinson built for Samuel Cabot, circa 1912. The temperature inside this shed often exceeded 100°F during freezing winter days.

Green's Ready-Built Homes of Rockford, Illinois, developed this prefabricated solar home for the postwar housing market.

Chapter 15
An American Revival

For two decades after Atkinson published his book, solar architecture lay dormant in the United States. But the progress made in Europe during the years after World War I helped to stimulate a new wave of interest. A number of European reports on proper building orientation began to appear in the English language. The most influential of these was a study by the Royal Institute of British Architects (R.I.B.A.) in 1931-32. The Institute published a clear, easy-to-follow reference manual on the sun's daily and seasonal movements and the number of hours a day that its rays would strike windows facing in various directions. The Institute also developed a simple device called a *heliodon* that helped planners determine the sun's effects on a building while it was still in the planning stages. By mounting a model of the proposed structure on a rotating drawing board and using a fixed light source to simulate the sun, designers could easily see what kind of solar exposure a building would have.

The charts and planning tools quickly caught on in Europe and soon found their way into the studios of American urban planners like Henry Wright of Columbia University. Wright advocated using this information to determine how to take best advantage of the sun's heat. Between 1934 and his death in 1936, he published many articles about sun orientation and solar communities like Neubühl in Switzerland. These essays helped convince other design professionals of the importance of solar architecture.

Another piece of convincing evidence appeared in 1934. While experimenting with heat-absorbing glass, invented to keep buildings cool in summer, the American Society of Heating and Ventilating Engineers (ASHVE) accidentally made an important discovery. They built two identical test structures, one with a south window made of the heat-absorbing glass and the other with a south window of ordinary glass. The building with the ordinary window, which transmitted the most winter sunshine, took nine percent less electricity to heat. Astonished, the president of ASHVE remarked:

> It has long been the custom to allow for extra heating on the north sides of buildings
> . . . because of the cold winds from the north. I think we are coming to realize that
> the provision of extra heat on the north side is not so much due to this effect as to
> the absence of sun heat!

Two years after this ASHVE study, the Pierce Foundation commissioned Henry N. Wright, son of the late urban planner Henry Wright, to determine exactly how much solar heat a building in New York City could gain during different seasons with its windows located on various sides. At that time, large amounts of glass were being used increasingly in building construction, according to Wright, often with "disastrous results." After World War I, factories began to mass-produce large, high-quality sheets of glass at reasonable prices. In addition, the development of structural steel eliminated the need for wood or concrete walls to provide support. Now that a steel skeleton could simply be covered with a glass skin, architects latched onto the idea of expansive windows and glass walls for purely aesthetic purposes. But indiscriminate use of glass on the north side of buildings led to cold rooms in winter—and much greater fuel consumption. If the east or west walls had large glass windows, too much sun came in during the summer.

Wright tackled the relationship between window orientation and building heat by

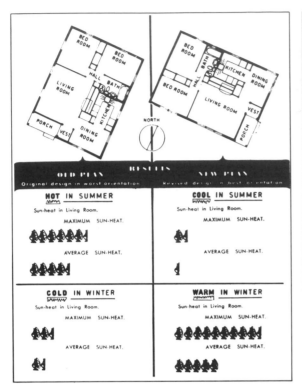

Top, left: Wright's chart relating house orientation to interior comfort.

Top, right: George Fred Keck, a Chicago architect who began an American Renaissance in solar architecture during the 1930's.

Above: The "House of Tomorrow" designed by Keck for the 1933 World's Fair.

applying weather information for New York City to the information published by
the R.I.B.A. on solar exposure. As Atkinson, Unwin and many others on both sides
of the Atlantic had "discovered" before, a southern orientation was found to be the
best for winter heating and summer comfort. Publishing his results in the influential
magazine *Architectural Forum* in 1938, Wright reached thousands of readers with
this message.

A Boom Begins in Chicago

While these theoretical studies were going on, a Chicago architect named George
Fred Keck began putting solar architecture into practice. He started his odyssey in
1932 while working on the futuristic "House of Tomorrow," which he had been
commissioned to build for the Chicago World's Fair. In an effort to make a lasting
mark, the ambitious young architect designed a three-story, twelve-sided building
with 90 percent of its walls made of plate glass. Keck had never heard of Greek and
Roman solar architecture, nor was he aware of Atkinson's work or the modern solar
communities in Europe. Because he had selected glass for form rather than function, it faced in all directions.

One winter day Keck had an experience that convinced him once and for all that
glass could help heat homes. He happened to visit the site of his project and noticed
the following:

> The workmen were inside finishing it up. This was in January or February . . . [and]
> the sun was shining very brightly. There were about half a dozen workmen in their
> shirts, without coats on. Yet there was no artificial heat, [for] they hadn't installed
> the heating plant. It was below zero outside and the men were working with just their
> shirts on and were comfortable in the house—it was the heat of the sun!

Keck was amazed by the heat that had accumulated in the glass house. He soon
began to think seriously of using glass to trap solar energy for home heating.
Unaware of previous work, he had his brother William obtain from the Weather
Bureau information on the sun's course during the year, the percentage of sunny
days, wind direction and other essential data. From these he deduced that a
south-facing glass wall with overhangs was best for year-round comfort.

Without access to institutional funding, however, Keck could only try out his
theory on the homes he normally designed for private individuals. But he had to be
cautious. Conventional wisdom taught that large glass windows would result in
great heat losses during winter. Hence, his approach was gradual. "Each year we
would build a small house for somebody," he recalled, and "each year we tried to
orient it [toward the sun] and open more and more glass to the south."

The introduction in 1935 of double-pane glass by the Libbey-Owens-Ford Glass
Company helped Keck retain more heat in his solar-oriented homes. These windows were made of two sheets of glass hermetically sealed together with a half-inch
air space between them. They cut 50 percent of the heat losses normally experienced with single-pane glass windows. He now began using double-pane windows
in all his homes. Overhangs prevented these windows from causing stifling conditions inside as a result of the hot summer sun.

Keck periodically visited the homes he designed, recording their fuel consump-

The first "solar home" of modern times, built for Howard Sloan in Glenview, Illinois.

tion and the owners' reactions. He found that the more glass he used on the south wall, the more solar heat the house gained. Buttressing his convictions, greenhouse owners in the area told him that their furnaces were shut down from dawn to dusk—even on the coldest days—if the sun was shining brightly.

After seven years of experimentation, Keck finally felt confident enough to expose the entire south side of a house to the sun. His opportunity came in 1940 when he designed a house for an old friend, Howard Sloan, a Chicago real estate developer. The house Keck designed was long and narrow. Its main rooms faced the all-glass south side, and overhangs kept out the direct rays of the summer sun. Sloan enjoyed this house and decided to test the sales potential of the bold, "modern" style. He explained how he promoted the house:

> I planned to exhibit it to the public. As it is customary to give some name to a model house and since the sun was used as a means of heating, I chose the word "solar."

A local newspaper called this house a "solar home," coining the term that has since been applied to many different sun-heated homes.

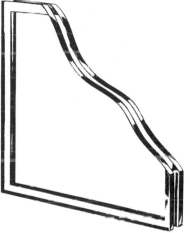

Left: Howard Sloan, Chicago real estate developer who built Solar Park.

Below: Thermopane *insulating glass, developed in 1935 by Libbey-Owens-Ford.*

Sloan was a good promoter. From the opening of the house on September 8 until Christmas of that year, over 5,000 people paid a dime each for the privilege of seeing the new "solar home." The national media were quick to pick up the story. *Business Week* dubbed the house design "the newest threat to domestic fuels." The professional community praised Keck's work. Many planners and architects had been recently exposed to the notion of solar heating and were becoming familiar with the diagrams and data generated by the R.I.B.A., Wright and others. Keck's prototype was the example needed to convince them to go solar. The surge of public support added to the favorable atmosphere. More than 300 people who had visited the Sloan house stated an interest in owning or building a solar home.

Sloan wasted no time in capitalizing on this interest. In 1941 he built a small housing development in the north Chicago suburb of Glenview—with half the homes of conventional design and the other half solar. The solar homes outsold the others by an overwhelming margin. For his next project Sloan built a development of 30 houses called Solar Park—the first completely sun-oriented residential community in the modern United States. Meanwhile Keck and a number of other architects were building solar-oriented structures that ranged from single-family homes to public schools and hospitals.

The Skeptics Take a Look

Not everybody agreed that solar homes used less energy. Skeptical engineers sought more exact figures on how much energy a solar home saved. In October

Two south-facing homes in Sloan's "Solar Park," the first solar-oriented subdivision in the United States.

1941, Libbey-Owens-Ford, developers of the double-pane window glass used in all of Keck's homes, sponsored a year-long study of solar heating. Tests were conducted by the Illinois Institute of Technology under the guidance of Professor James C. Peebles and William Knopf, a graduate student.

The two engineers tested the solar performance of a typical solar home built by Keck and belonging to Dr. and Mrs. Hugh Duncan of Flossmore, Illinois. Knopf and Peebles wanted to study conditions in a real-life situation rather than set up an artificial laboratory experiment. That winter even they were surprised at the large amounts of heat that accumulated inside the house. One day when the outdoor temperature rose to only $-5°F$ the house warmed up so quickly that by 8:30 in the morning the thermostat shut down the furnace, which didn't go back on until about 8:30 that evening. The room temperature often reached 85°F and occasionally the occupants had to open the doors and windows for some cool air. Such periodic ventilation made it difficult to calculate how much solar heat had actually collected inside the house. Occasional overheating of the house was partly due to the conventional heating system used: hot water pipes beneath a concrete floor that radiated heat into the rooms. Because it took a long while for it to conduct through the massive floor slab, heat continued to be released long after the thermostat had shut off the furnace. This added room heat made it uncomfortably warm inside the house. And when the windows were opened, more heat often escaped than was desirable. Knopf and Peebles reached the following conclusion:

> A house of similar design equipped with a heating system better adapted to fully utilize the available solar heat input should show a substantially lower heating cost than that recorded in this test.

Other problems that made the test results unclear were the random opening of doors and the loss of heat through cracks around doors and windows that had not

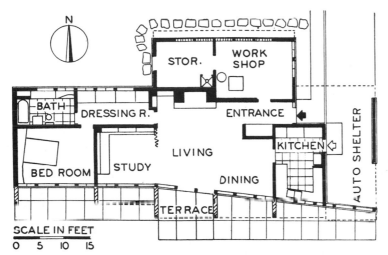

Southwest view of the Duncan house in Flossmore, Illinois. To evaluate the merits of the emerging solar architecture, a team from the Illinois Institute of Technology tested this solar house during the winter of 1941-42.

been caulked as planned. The editors of *Architectural Forum* who analyzed the Institute's report estimated that the sun provided between 7 and 18 percent of the home's heat. But Knopf and Peebles would only offer a conservative statement:

> The preponderance of evidence indicates that the solar heat input in the test house offset most, and probably all, of the heat losses through the extra window areas and kept the total heating cost at a reasonably low figure in spite of the loss of heat through excessive infiltration and frequently opened windows and doors.

Although the researchers were unable to determine the precise solar heat gain of the Duncan house, it was apparent that a solar house with few cracks to cause heat leakage and with a more compatible heating system would collect a substantial amount of heat from the sun. And there was another energy-saving dividend. The large south windows created an ambience of total illumination. At no time during

188

Top: Winter sunlight streaming into the living room and study kept the Duncan home warm even during sub-zero weather.

Right: No direct sunlight entered the Duncan home during the summer.

the day "was artificial light required in any part of the house, no matter how remote from the windows."

Solar Homes Stir the Public Imagination

People who owned solar houses didn't need complicated technical reports to tell them about the advantages of solar heating. With wartime rationing making oil and gas conservation mandatory, homeowners appreciated staying warm in winter by the sun's heat. Arthur Carstens, for example, wrote a letter telling Keck of his great satisfaction:

> Sometimes I thought you were stretching a point when you talked about the advantages of solar heat. I am now satisfied that you were not, and the records of our fuel dealer can well support my statement. We could easily have gotten along on one half the oil which the ration board allocated us.

The J.L. Bennetts of Barrington, Illinois, said they saved over $100 a year, or 30 percent of the fuel bill of their neighbors who were living in a house of comparable size. Many residents claimed they saved as much as 38 percent on fuel. Institutions saved a lot of money, too. For instance, it cost $1,500 less to heat the solar-oriented Lake County Sanitarium than a conventional building of similar size.

While solar homeowners could gloat, others had to wait until the war's end to share the benefits. Wartime restrictions on the use of building materials had halted almost all new construction. But many articles about solar homes appeared in the professional journals and popular magazines of the day. *Newsweek* wrote a favorable summary of Knopf and Peebles' report, stating "One way for Americans to hedge against future fuel shortages would be to build more solar homes like that of Mr. and Mrs. Hugh Duncan." *House Beautiful* agreed with this high appraisal: "No idea of the last 30 years has so fired the imagination of the American public as the one of letting old sol reduce the winter fuel bill." And people could enjoy the cheery warmth of a sun-drenched house without having to settle for the stark, bold "modern" lines of a Keck house. According to an article in *House Beautiful,* "Any clever architect can give his client both solar heating and the style of his heart's desire." The magazine frequently ran stories on solar homes of all types, ranging from Colonials to Cape Cods. In the periodical's popular Home Planner's Study Course, readers discovered how their old house could be remodeled into a solar home if the main rooms already faced south.

Other magazines took up the cause of modern architecture and championed solar features as an integral part of the new look. In 1944, *Ladies Home Journal*—a leading advocate of the innovative house designs of the day—commissioned a dozen of the nation's leading architects to draw up plans for "small but really adequate houses which would dramatize the advantages of modern planning." Among the notables in the project were Phillip Johnson and Frank Lloyd Wright. In almost every case, the principles of solar architecture were used, and month after month the *Journal* featured articles on these homes. With its readership in the millions, the public response was enormous. Thousands sent in comments, such as the couple who wrote, "Because of your article describing the solar house, we have changed our ideas in respect to what we wanted in a home. We have visions of nothing but one of these homes after the war.'' Given prominent play on the pages of popular periodicals, reactions like these had a tremendous influence on those idle developers who were gearing up for the inevitable building boom that would follow the war's end.

Solar Building Takes Off

The housing market exploded after the war ended. In 1944, only 144,000 new homes had been built; by 1946 the figure hit a million a year. Many manufacturers in the prefabricated home industry, such as Green's Ready-Built Homes of Rockford, Illinois, and Wickes, Inc. of Camden, New Jersey, adopted solar designs. New companies using mass-assembly techniques, such as Solaray Homes of Boston, also entered the field. One of Solaray's developments in South Natick, Massachusetts, consisted of 16 homes selling for about $10,000 each. The houses were individually site-planned, with large bay windows on the south side and eaves to

Advertisement for prefabricated solar homes from a 1945 issue of Popular Science.

YOUR SOLAR HOME IS ALL WRAPPED UP

Pretty as a picture is this view from living room of prefabricated Solar Home. Complete in package form, this home will be delivered to you by truck

block the summer sun. Solar developments built by other companies ranged in size from 15 units to well over a hundred. In the Chicago suburb of Northbrook, Howard Sloan built 18 two and three-bedroom houses that sold for $15,000 each. The homes on the north side of the street fronted south, while those on the south side had their large glass windows facing the backyard. In addition to these developments, many individual architects designed custom solar houses all over the country.

The renaissance of solar architecture renewed the controversy among engineers about the effectiveness of south glass windows. As one engineer put it,

> Substantial reductions in seasonal heating costs are said to have been found from some installations while results of an opposite order are attributed to other installations. Where, in this confusion of claim, opinion and estimate, does the truth lie?

In an attempt to answer this and other questions, Libbey-Owens-Ford sponsored a second study of solar house heating. It was conducted in 1945-47 by Dr. F.W. Hutchinson, Professor of Mechanical Engineering at Purdue University. Dr. Hutchinson found that one of the primary reasons for varying degrees of success was the lack of reliable information on the optimum ratio of glass area to floor area. This ratio should vary with climate; if the glass area were too large for a particular climate, the house would overheat and solar heat would be wasted. If the windows

A prefabricated solar house offered by Wickes, Inc. of Camden, New Jersey.

were too small, insufficient solar heat would be obtained. Hutchinson developed a set of tables that allowed architects to make more accurate estimates of the proper amount of south-facing glass.

Controlling the Solar Heat

Another valuable outgrowth of Hutchinson's study was his demonstration that a solar home has much greater temperature swings than an ordinary home—it warms

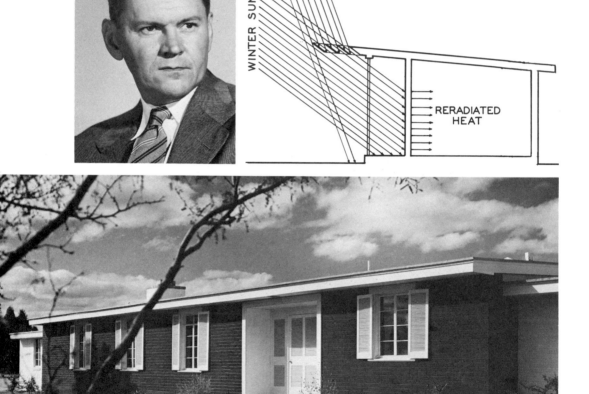

Top, left: Arthur Brown, architect. He discovered the solar heat-storage potential of masonry walls while working on a house (photo above) with a blackened south wall.

Top, right: Brown's diagram of the heat flow in the Tucson solar house (photo opposite page).

up quickly during the day due to the accumulation of solar heat, and cools off rapidly at night due to excessive heat losses through the windows. Some architects realized it was advisable to protect windows at night with heavy drapes or shutters so that less heat would be lost. Furthermore, it made sense to develop a storage system to capture the excess solar energy during the day for use at night.

Even before Hutchinson's study was complete, an Arizona architect named Arthur Brown developed a heat storage system that allowed solar heat to be used at night. In 1945 Brown built a long, narrow home in Tucson with the broad sides of the building facing north and south. For aesthetic reasons the couple who owned the house wanted one of the outside walls painted black. Knowing that a black wall on the east or west would get too hot in the torrid Arizona summer, he suggested painting the south wall. One winter day while inspecting the house, Brown was startled by the amount of heat emanating from this black wall. "I could feel it five feet away," he remembered, "and I thought that the next time we do a house, we'll paint the wall inside the hall black so that we won't lose that heat."

Brown's next house was a long, narrow building with a south wall made entirely of glass. A glazed sun porch and dining room occupied the south side of the house, and a black concrete wall ran down its center. On the north side of this dividing wall were the bedrooms, living room and kitchen. During clear winter days, sunlight struck the black central wall and solar heat penetrated the concrete. Brown had estimated that heat moves through concrete at the rate of one inch an hour. So he built the wall eight inches thick, and by evening the solar heat was just beginning to

Southeast view of Arthur Brown's solar home in Tucson, Arizona. A masonry wall inside this house stored solar heat for nighttime use.

Interior view of the Tucson solar home on a sunny winter day. Sunlight streaming in through the south windows warms the masonry wall at left.

penetrate the wall and radiate from the back side into the rooms in the northern part of the house. By morning, the black concrete wall would be cool, ready for another cycle. He also recommended that curtains be drawn behind the glass soon after sunset to retain the solar heat. Although this heat storage system was never tested quantitatively, it definitely increased the sun's share in heating the home.

In the mid-forties another improvement appeared—adjustable canvas awnings for the south windows, similar to the type used in Europe for some time. This device was important, as Keck explained, because "in springtime the temperature can still be cold when the sun is well past the equinox (March 21)." Stationary overhangs might keep out the September sun when it was undesirable, but they would also block the sun in March, when it followed the same path through the sky. Adjustable overhangs solved the problem.

The Decline of Solar Architecture

Despite these improvements, solar building began to falter by the late 1940's, partly because the wartime conservation ethic had swiftly faded. With energy becoming ever cheaper, few cared if the sun could help cut heating bills. People were more concerned with the initial cost of a house. A solar house—due to the extra expense of double-pane glass and individualized siting—had a price tag up to 10 percent higher than that of a comparable conventional house. In the extremely competitive housing market, these higher first costs drove customers away or led to lower profit margins if developers reduced prices to attract buyers. Companies that mass-produced solar homes encountered other problems. Most pre-fabs, including those that were solar, were built with special lightweight wall panels that helped keep construction costs down. These panels did not meet the antiquated building codes of most cities, however, and many companies in the business folded.

The reputation of solar homes also took a beating. The look of large glass windows was attractive to many people, but a lot of them didn't realize that the purpose of these windows was functional—not merely aesthetic. Consequently, the public readily bought homes designed with improperly oriented windows. One Long Island developer built thousands of homes with large, randomly-oriented windows. Such misapplications of solar design often led to disastrous results— overheating in summer and huge winter heat losses. Many times these structures were called solar homes, leading to a muddling of the term.

Despite these snags, sun-oriented homes continued to be popular for a time among the wealthy, who could afford to hire an architect well-versed in the techniques of solar heating and natural cooling. Custom-designed solar homes flourished from Maine to California and from Texas to Canada during the late 1940's. Eventually, however, the increased popularity of mechanical heating and ventilation systems—compounded by constantly falling fuel prices—led to an almost universal lack of interest in solar architecture by the late 1950's.

Southwest view of M.I.T.'s fourth solar house in Lexington, Massachusetts. Planned as a prototype for mass construction, it obtained half of its winter heat from the sun.

Chapter 16
Solar Collectors
for
House Heating

During the 1940's, a number of engineers and scientists tried a completely different approach to solar house heating. They decided to use solar collectors resembling a hot box or the Sun Coil for this task. The low-temperature solar heat from these collectors was well matched to the heating needs of a home. Many of these experimental homes also had a heat storage unit separated from the solar collector. The solar heat collected could then be piped to a massive heat storage volume located inside the house and insulated from its surroundings. This way, the heat would be available as needed—whether at night or on cloudy days. The approach resembled that taken by William J. Bailey when he split the functions of solar heat collection and hot water storage in the Day and Night solar water heater. Its advocates thought that such an integrated system of collectors and storage would use the available solar energy more efficiently. Excess solar heat could be stored for later use, rather than overheating the house or escaping through open windows. And with thermostats controlling the flow of heat through the house, its occupants would remain more comfortable.

Morse's Solar Air-Heater

The use of solar collectors for house heating was not a completely new idea. The earliest recorded instance dates back to the 1880's when Edward S. Morse, a botanist and ethnologist of worldwide reputation and Lecturer at the Essex Institute in Salem, Massachusetts, used a hot box for this purpose. He observed that when dark curtains are closed behind a sun-drenched window, they become very hot and warm air currents develop between the glass and curtains. So why not use the same principle to heat a room or a house? Morse built a first, experimental apparatus to test his theory. This device was little more than a hot box attached to the south wall of a building, with openings to permit outdoor air to enter the box and solar-heated air to escape through a vent into the rooms.

Early in 1882, Morse's first solar air heater was installed at the Peabody Museum in Salem. *Scientific American* called it "an ingenious arrangement for utilizing heat in the sun's rays in warming our houses . . . [yet] so simple and self-contained that one wonders that it has not always been in use." Along the outer wall of the museum stood a hot box approximately 13 feet high and 4 feet wide. The wall formed the back side of the box, and on the front a glass cover faced the sun. In the middle of the box, several inches away from the wall, Morse put a sheet of black corrugated iron to absorb the sun's rays. Cold outdoor air entered through an opening at the bottom of the box, flowed around the iron sheet, and picked up solar heat from it. The warmer, lighter air then rose up to an opening at the top of the box and into the room. But the heater was much too small for the 4,000 square-foot hall, and it had little effect on the room temperature. However, Morse noted that the air temperature increased by at least 26°F after passing through this hot box.

Morse decided to make some design improvements and build another test collector for the study of his house. He replaced the corrugated iron with slate and placed the glass perpendicular to the ground, so that it resembled a large window. Cold air could be drawn into this solar heater from the outdoors as before, or room air could enter at floor level. Either way, the air rose through the box as it warmed, and entered the room near the ceiling. The temperature of room air entering the box

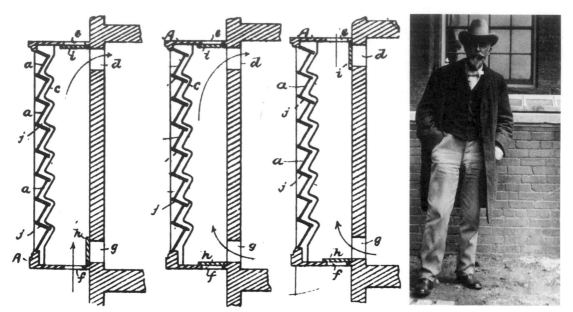

Above: Professor Edward S. Morse (right). Patent drawings of his first solar heater, 1881 (left), showing three different modes of operation.

Opposite page: Morse's home in Salem, Massachusetts (bottom). His second solar heater is just barely visible behind the bay window at the left of the photo. Cross-section of the solar air heater on Morse's study (top left). Sun-warmed air rose behind the slate and passed into the room through the upper duct. Interior view of the study (top right).

rose from 64°F to 87°F on sunny days; this boost was enough to keep the study warm. The slate heater worked much better than the one in the Peabody Museum because the ratio of heater size to room size was much more favorable. Morse emphasized that his invention could only serve as an auxiliary heater, because there were too many overcast days during the Salem winter—not to mention the lack of sunshine at night. Still, the solar heater had many advantages—cutting down on the dust, cinders and fire danger from coal or wood-burning stoves.

Morse built his last and largest solar heater for the Boston Athenaeum—the place where Alexander Graham Bell had first publicly demonstrated his telephone several years before. This solar heater was a larger version of the one designed for his study, measuring 42 feet high and 6½ feet wide, for a heating surface of 273 square feet. Unfortunately, the warm air entering the room from the heater lost much of its heat through a large skylight in the ceiling. Nevertheless, the device helped to save the Athenaeum 25-50 pounds of coal a day during winter.

Several newspapers picked up the story of Morse's attempts, but much to his chagrin they grossly exaggerated his success. The *Troy Daily Times* called him "a man who thinks he has solved the problem of house heating." The *News of the World* wrote: "It is said that a certain professor down in New England . . . warms his home during the daytime . . . by the heat of the sun alone." The *Salem Gazette* credited Morse with "finding a way to outwit the coal combine." He even received

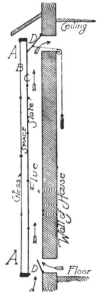

a letter from the Los Angeles office of the Climax Solar Water Heater Company, asking him to let them know "whether [he] intends to put them [the air heaters] on the market, and what arrangement I [the general manager] can make . . . to handle them."

Obviously upset by the public misinterpretation of his work, Morse sarcastically remarked that he expected the next news story to describe his house in the following manner:

> A visit to the unique structure in an intensely cold snap with the thermometer at zero revealed the male members of the family roaming round in pajamas, having just gathered a large crop of oranges, the ladies swinging in hammocks and fanning themselves while fronds of gigantic tree ferns formed canopies in every room.

Although Morse parodied the inept reporting of the news media, he sincerely believed that his solar air heater would have useful and important applications. He knew that solar energy could help save fuel, especially in energy-short regions:

> The results were sufficiently satisfactory to show that in regions such as the great [American] Southwest where fuel is scarce, temperatures often low, and [there are] many sunny days, the invention could be made of great service.

But despite Morse's prognosis and the extensive publicity he received, his idea fell by the wayside for another half a century.

The First M.I.T. Solar House

In 1938, engineers at the Massachusetts Institute of Technology (M.I.T.) began two decades of research on the use of solar collectors for house heating. A grant of $650,000 bankrolled this extensive study. The money was a gift from Godfrey Lowell Cabot, a wealthy Bostonian then described by *Time Magazine* as a man of 77 who in his old age "broods much about the vast stores of energy in sunlight which man does not utilize."

Hoyt Hottel, a professor of chemical engineering at M.I.T., became director of this research program. He was assisted by Byron Woertz, a graduate student, and others at the Institute. Hottel was aware of the many prior developments in solar energy; his objective was to test their home-heating potential scientifically. He chose first to study intensively the flat-plate collector—the type of solar collector successfully used to heat domestic hot water supplies in California and Florida. He favored this device because of its simplicity, its popularity in Florida at the time, and its potential for economical home heating. Hottel also liked the fact that its heat transfer and storage medium, water, has a very high capacity for absorption and retention of heat.

The research staff set out to conduct a rigorous scientific study of the technical and economic feasibility of heating a building exclusively with solar energy. Their first task was to construct a small building that would serve as the experimental laboratory. Unfortunately, the architect and builder used an uncorrected magnetic compass to determine geographic north—which put the house about 7° off a true-south orientation. "Hottel was about ready to bite into a nail" when he learned of this blunder, said Woertz.

A workman installs flat-plate collectors on the first M.I.T. solar house, 1939.

When the structure was finished, 14 flat-plate collectors were installed on the south side of the roof at a tilt angle of 30° from the horizontal. The collectors occupied a total of 408 square feet—almost the entire south side of the roof. Each collector consisted of a shallow plywood box with six parallel copper tubes inside it soldered to a blackened copper plate. This configuration was very similar to that used by William J. Bailey in his earliest solar water heaters. On top of the box there were three glass covers separated by air spaces, and on the bottom were over five inches of rock wool insulation. Hot water was pumped through the copper tubes to the roof peak and then down to a 17,400-gallon storage tank that took up the entire basement. The insulation covering this huge tank averaged a little over two feet thick. To heat the building, cool air was drawn from the rooms by fans and blown over the hot tank. The warmed air then circulated back into the rooms. In this "active" solar heating system, an external source of electrical power was required to activate pumps and fans so that heat could be moved from the collectors to the storage tank and from storage to the rooms. Unlike Keck's solar homes and Morse's solar air-heater, this system could not function without power.

The solar heating system began operating in midsummer, 1940. Hot water was pumped from the rooftop collectors to the storage tank continuously. During winter, whenever water in the collectors became cooler than water in the storage tank, a control box shut down the system and allowed the water in the collectors to drain back to the tank. That way heat was not lost from the storage tank, and there

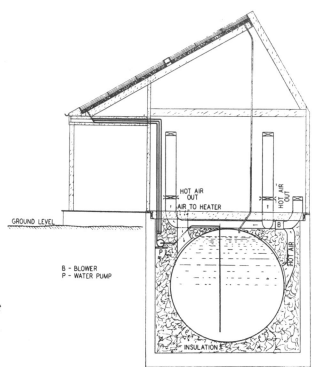

*Cross-section of M.I.T.'s first
house. Solar heat from the
rooftop collectors was stored
in a 17,400-gallon water tank
that filled the basement.*

was no danger of water freezing in the collector and splitting the copper pipes.

From a technical viewpoint, the results of the experiment were generally positive. By means of solar collectors alone, the laboratory remained at a steady 72°F throughout the winter. However, the results of the economic analysis were clearly unfavorable. Relying completely on solar energy meant year-round collection, which required a large, costly array of collectors and a massive, heavily-insulated storage tank. But aside from demonstrating technical feasibility, the experiment yielded important data. The report on this test written by Hottel and Woertz is still regarded as a classic in the field. The engineers isolated the principal factors affecting the performance of a solar collector—the tilt angle of the collector, the light transmission of the glass covers, heat loss through the bottom of the plywood box, the type of absorber plate used, and several other important factors. The M.I.T. study demonstrated that the efficiency of a flat-plate collector drops as the temperature difference between the absorber plate and the outside air gets larger. Efficiency also drops if more than three glass covers are used. One surprising discovery they made was that airborne dirt and soot have little effect on collector performance.

A Solar Hot-Air System

Instead of using water to transport and store solar heat from the rooftop collectors, some engineers chose to develop systems relying on solar-heated air. This approach pioneered by Morse in the late nineteenth-century had long been forgotten.

*Dr. George Löf (foreground) and his research team unveil their prototype solar
air-heating collector in 1944.*

His collectors operated by natural convection of solar-heated air but could not store
solar heat for nighttime use. By the middle of the 1940's, two people who had been
involved in the first M.I.T. experiment began to work with full-scale solar hot-air
systems for house heating.

Dr. George Löf, an associate professor of chemical engineering at the University
of Colorado, had done his graduate work under Hoyt Hottel's guidance and was
familiar with the M.I.T. experiment. In 1943, he became director of a project to
devise an effective, economical solar heating system. Ironically, while World War II
had interrupted the M.I.T. experiments (everyone on the staff had been reassigned
to war research), it led to the commencement of this new solar project in Colorado.
The War Production Board funded the study, said Löf, because they were concerned about the possible fuel shortages due to heavy military requirements.

Löf quickly assembled a group of aides and set to work. He decided not to follow
the M.I.T. approach because:

> Probably the main reason . . . [those at M.I.T.] went into the use of water was that
> solar water heaters had been used for many years and that was the easiest route—
> just make them bigger and use them to heat houses. But our rationale was that
> we shouldn't be constrained or biased by how solar energy had been used before.

Löf believed that a hot-air system was the logical choice because most homes at the
time were heated by hot air; existing heating systems could easily be combined with
solar air-heating systems. And in cold climates where most Americans then lived,

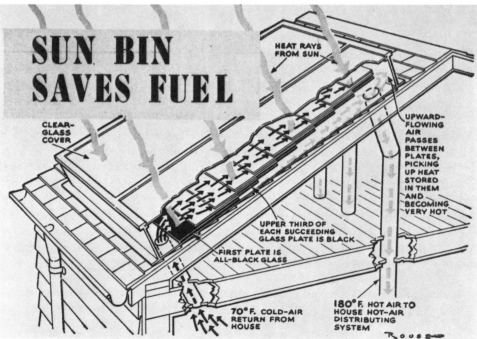

SUN BIN SAVES FUEL

HEAT RAYS FROM SUN

CLEAR-GLASS COVER

UPWARD-FLOWING AIR PASSES BETWEEN PLATES, PICKING UP HEAT STORED IN THEM AND BECOMING VERY HOT

UPPER THIRD OF EACH SUCCEEDING GLASS PLATE IS BLACK

FIRST PLATE IS ALL-BLACK GLASS

70° F. COLD-AIR RETURN FROM HOUSE

180° F. HOT AIR TO HOUSE HOT-AIR DISTRIBUTING SYSTEM

Top: Dr. George Löf at his solar-heated bungalow in Boulder, Colorado.

Above: Cut-away view of Löf's glass-plate collector.

During the second winter of testing, crushed-rock heat storage improved the performance of Löf's solar-heating system.

solar collectors using air would have no problems of cracked pipes due to freezing, and corrosion and leakage would be eliminated. Water does have distinct advantages, however. It carries far more heat than an equal volume of air, and a water heating system requires narrower piping and less storage volume—thereby reducing initial costs.

The Colorado team sought to cut the system costs by developing an all-glass collector. Two layers of glass covered the collector box, inside of which lay several glass plates that were partially overlapped—somewhat like the shingles on a roof. The upper surface of the overlapping plates was painted black to absorb the sun's rays. Preliminary tests with this collector were encouraging.

But when it was time to put this technique into full operation, the research team was confronted with somewhat of a dilemma. The war had halted new construction, so they had to find an existing home in which to test the system. Löf volunteered his cozy, six-room house in Boulder—which was then heated by a gas-fed hot air system. Over 460 square feet of collectors went up on the roof of the 1,000 square-foot bungalow. Air from the rooms entered the solar heating system through a duct in the roof. The air temperature rose by as much as 110°F before it circulated down to the furnace, where a fan pushed the warm air through the regular heating ducts. When solar heat was insufficient, the gas furnace took over. This was the first time that anyone had integrated an active solar-heating system with an existing house-heating system.

During the first winter, the sun provided about a quarter of the heat needed in the house. To improve this figure, Löf knew he needed a storage system so that solar heat could be supplied during cloudy days and at night. He considered various materials for absorbing heat from hot air—such as crushed granite, cinder blocks and hollow tiles. Rock was the cheapest option. Löf calculated that six tons in an insulated bed located beneath the floor would hold enough heat to keep the house from getting too cold overnight. The advantage of a larger storage holding enough heat for two or three days would be offset, he felt, by the higher initial cost of building it.

With the rock bed heat storage, the system now provided about a third of the home's heating needs the following winter. This test proved that an ordinary house could be refitted fairly easily with solar equipment. But the all-glass collector did not work out, because the overlapping plates tended to crack under heat stress. And when the Löfs were ready to sell their house after living with the solar heating system for several years, they felt compelled to remove the collectors. Löf told why:

> Nobody at that time was interested in using solar because fuel was so cheap, and if there were any maintenance problems, the new owners of the house would probably feel helpless because a typical heating and air conditioning installer wouldn't know much about the system.

But he had shown convincingly that the sun could provide a substantial heating boost even in regions like Colorado with icy cold winters.

Water-Wall Collectors

After World War II interrupted the M.I.T. project, it was not until 1947 that Hoyt Hottel and his team resumed their research. To develop a more economical system, they sought to combine in a single unit the functions of collecting, storing and distributing solar heat. They erected a wall of water containers behind vertical south-facing glass—a "water-wall" solar heating system. In this experiment, an array of one and five-gallon water cans painted black were stacked up just inside

South wall of the laboratory used by M.I.T. for its experiments with "water-wall" solar collectors. Insulating curtains are drawn in all but one unit.

a south wall made almost entirely of double-pane glass. Dr. Albert G. Dietz, an engineering professor involved with the project, explained how the system worked:

> The sun hit the cans of water through a couple of sheets of glass. The water got warm, and then that energy was transmitted to the interior of the house. It was obviously a much simpler system than the usual one where you have a flat-plate collector and either air or water to take the sun's energy from storage into the house.

The building used to test this water-wall approach was a long, narrow laboratory aligned east-west and facing south; it was divided into seven small cubicles, each with a slightly different solar heating system. The first cubicle had a large south window with no water cans behind it. In the other six cubicles cans of water absorbed the sun's rays, warmed up and transmitted this heat to the interior. To prevent overheating of the rooms, the engineers inserted a curtain separating the water cans from the interior in four of the cubicles. The curtain was raised whenever the interior temperature fell below 72°F, allowing solar heat into the room. In the

Inside the laboratory control room, a student monitors the instruments.

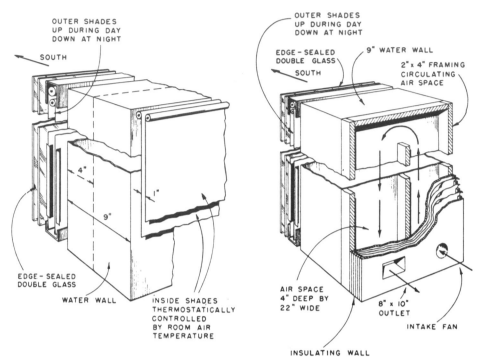

Schematic diagram of the two kinds of water walls.

remaining two test cells, a fan controlled by a thermostat blew room air past the water cans and back into the interior—similar to the approach used in the first M.I.T. house.

All of the water-wall systems exhibited one major flaw—a lot of heat escaped through the glass from the warm water cans sitting right next to it. To reduce this heat loss, the engineers installed aluminum curtains between the cans and the glass; these curtains were drawn as soon as the sun went down. Even so, heat losses ranged from 71 to 84 percent of the solar energy collected. Evidently these curtains failed to block warm air currents from reaching the cold window surfaces.

Despite these heat losses, the water walls were still able to supply between 38 and 48 percent of the heating needs of each cubicle. Better insulation between the water cans and the glass might have made this water-wall approach a successful house-heating option for the cold New England states. But the M.I.T. research team did not pursue the idea any further. Nor did they compare the cost of such a simple heating system with the cost of a more efficient but highly complex and expensive flat-plate collector system. It might have been wise to sacrifice collector efficiency for lower costs and greater reliability.

Combining Collectors with Solar Architecture

Disappointed by the large heat losses from the water wall, the M.I.T. engineers returned to an active system that pumped solar-heated water from rooftop collectors to a storage tank. Two important modifications of their first active system

A student family occupied M.I.T.'s third solar house, which used both flat-plate collectors and south-facing windows to collect solar heat.

brought down the cost. First of all, the design goal was no longer to heat a home entirely with solar energy. As Dietz observed,

> Since the collector is the most expensive part of an installation, it just isn't very economical in most places to have one large enough to take care of all your heating requirements under all conceivable conditions. The concept was that our solar collector should be designed to collect enough heat on a sunny day to carry a house through two or three sunless winter days. Anything beyond that, we would call on the auxiliary heater.

The auxiliary consisted of two small electrical immersion heaters placed in the hot water storage tank. Because of this auxiliary, less hot water needed to be stored for cloudy days and heating at night. So the collectors and tank could be built smaller—thereby saving money.

The second modification in this house was the use of large south windows, as in Keck's solar homes. The designers figured that these windows could trap enough heat on a sunny winter day to keep the house warm from morning until nightfall. All of the solar heat absorbed by the rooftop collectors during the day could then be stored in the tank for use at night.

This modified solar heating system was installed in the building that had been used for the second M.I.T. experiment. An A-frame roof was added and 16 collectors were installed on the south slope, tilting at 57° to the horizontal. These collectors resembled those used in the first experiment, except that they had only

The third M.I.T. house. Floor plan (above right). Sunlight streaming in the south windows (right) supplied 29 percent of the winter heat needed by the house.

two glass covers instead of three. Situated in the attic, the storage tank was less than one-tenth the size of the tank used in the first house. The long, cylindrical tank held 1,200 gallons of solar-heated water; it was insulated with rock wool ranging from 4 to 12 inches thick. Instead of air being blown over the storage tank and distributed to the rooms as in the first house, the hot water itself was circulated to radiant ceiling panels inside the house.

Because the engineers wanted to test the system under normal living conditions, an M.I.T. graduate student and his family moved into the house. Data collected during the first year of operation—from 1949 to 1950—showed that the system was strikingly successful. Solar energy provided almost three quarters of the home's heating requirements. The engineers were particularly surprised by the effectiveness of the large south windows. Hutchinson's tests at Purdue and the second M.I.T. study both seemed to contradict claims by Keck and occupants of his solar homes that south windows contributed substantially to house heating. In this third M.I.T. house, however, the windows supplied a whopping 29 percent of the heat needed inside—consistent with Keck's contention. The reasons for this discrepancy lay in the difference between sterile laboratory tests and actual living conditions. Furniture in the house absorbed some of the solar heat during the day and released it when the temperature cooled off. And the rental contract stipulated that the family had to close its curtains by 10 p.m. every evening to help keep the heat from escaping through the windows.

The collectors trapped 37 percent of the available solar energy and provided 44 percent of the home's heat. Body heat and waste heat from appliances contributed another 16 percent, leaving only 11 percent to be supplied by the electric heater. The solar heating system might have provided even more heat had the structure been specifically designed for it. But the house was converted from the earlier laboratory, and the storage tank had to be placed in the attic. Built long and narrow to fit the attic geometry, this tank suffered high heat losses through its large surface area. Had the tank been placed in a basement, much of the escaping heat would have risen into the house—helping to keep it warm. Instead, this heat just escaped into the atmosphere.

The solar heating system operated successfully from 1949 through 1953, when the experiment came to an abrupt end due to a series of mishaps caused by faulty wiring in the electric heater. When the wiring short-circuited and started to smoke, the fire department arrived and, as Hottel recalled,

> They got up on the roof, swung their axes, hit metal and insulation, decided this was not the place to open up, and then moved down and hit another. And they went along and opened up enough holes in the roof so the air could move through the attic. From nothing but smoldering, they built a fire for us and ruined the house.

The Dover House

At about the same time as the third M.I.T. experiment, a novel approach to solar house heating was being tested in Dover, Massachussetts—just outside of Boston. This project was organized by Dr. Maria Telkes, a Research Associate in Metallurgy at M.I.T. who had been working in the field of solar energy since 1945. She thought that storing heat in large tanks of water or tons of crushed rock was impractical. Only if a solar heating system provided *all* the house's heat, she reasoned, would it be economically feasible. But to store enough heat to carry a house through a week of cloudy winter days, huge volumes of water or crushed rock were needed—much too large for practical purposes.

Telkes sought to harness the capacity of certain materials that absorb great amounts of heat in the process of melting and release this "heat of fusion" when they cool and resolidify. For years, she had searched for an inexpensive, readily available material with a low melting point and a high heat of fusion. In 1946 Telkes found what seemed like the perfect substance—Glauber's salts or sodium sulfate decahydrate, commonly used in the manufacture of glass, paper and detergents. Over a particular temperature range, Glauber's salts could store seven times more heat than the same volume of water; its melting point was about 90°F, and at the time it cost only $8.50 a ton.

Dr. Telkes discussed Glauber's salts with Hoyt Hottel, who was also looking for a better way to store the sun's heat. He enthusiastically organized a group of graduate students to test the substance. But here an unexpected difficulty arose. According to Albert Dietz, one of Hottel's associates:

> The principal problem was simply this—when you first melt the salts they separate into two phases, the heavier sodium sulfate settling to the bottom and the saturated water solution on top. Then you'd have the problem of remixing them during any subsequent cycle. To get full efficiency, they should be remixed during the freezing cycle. That is the problem—how do you do that?

Unless the two stratified solutions could be remixed, the solidification would be only partial—and the heat recovery incomplete. At the laboratory they tried to control the stratification by using containers of various sorts, but all to no avail.

Hottel thought it premature to use Glauber's salts in a solar house without further testing. But Dr. Telkes was not satisfied with the way the experiments had been run. Disagreeing with Hottel, she found outside support to build a solar-heated house

Dr. Maria Telkes (left) and Eleanor Raymond—the scientist and the architect who tried for 100 percent solar heating in the Dover house behind them.

using Glauber's salts for heat storage. She announced that this house would be the first occupied residence to be completely solar heated. The idea interested Amelia Peabody, a wealthy Bostonian who agreed to finance the project. She asked her architect Eleanor Raymond to draw up the plans for a solar-heated house to be built on her estate in nearby Dover. This house thus became an "exclusively feminine project," as the *Saturday Evening Post* noted in 1949.

Rather than mounting the collectors on a sloping roof, Dr. Telkes stood them upright to form the south wall of the second story. This bank of 18 collectors ran the full 75-foot length of the house, which was long and narrow—only one room deep—to accommodate the necessary collector area. Fans blew air across the black iron absorber plates in these collectors, down through ducts and past cans of Glauber's salts located between the walls of adjacent rooms. The salts absorbed solar heat from the passing air stream and melted. When the rooms cooled down in the evening, another set of fans circulated room air past the cans of Glauber's salts, which released some of their heat to the airstream, warming the house.

But exactly how well the salts worked is not very clear. Did the salts actually melt and solidify? Or did they merely store heat by getting warmer—as would rock, water, and any other substance? This uncertainty occurred because Telkes used an enormous amount of the salts—470 cubic feet—that could store plenty of heat without ever melting. Such a large volume of salts was used because she designed for a seven-day storage capacity and had not installed any auxiliary heating system.

Unfortunately, the weather refused to cooperate. Right after Dr. Anthony

South face of the Dover house, with the Nemethy family gathered in front. Solar air-heating collectors spanned the entire second-story wall.

Nemethy, his wife Esther and their three-year-old son Andrew moved into the house on Christmas Eve, 1948, they were confronted with eleven cold, sunless days. "The solar heating system was exhausted," recalled Mrs. Nemethy. "As we didn't have any electric or oil heating or anything, we just were practically without heat." But during the remainder of that winter, which was milder than usual, the house stayed comfortable without any need for backup heating.

For two and a half winters, the Dover House functioned fairly well. Only when there was a week-or-longer stretch of cloudy winter days did the solar heating system fail to keep the house warm. During the third winter, however, the Glauber's salts deteriorated. Some cans began to leak, and the salts separated into two distinct layers—solid and liquid. Hence, the salts could not release the solar heat they had absorbed. Esther Nemethy recalled that winter with shivers:

> Both I and my son had very bad colds and there was a snow storm and there was no heating in the house. I called up Mrs. Peabody and said, "I'm sorry, we love Dover, we love you, we love the house, but I'd rather move out from here unless you install electric heaters or do something!"

After visiting the Nemethys and seeing how cold the house could get, Amelia Peabody installed small electric heaters in the rooms. And in 1953, she decided to install an oil heating system and remove the solar. As the tenants now needed extra room anyway, she took out the collectors and integrated the second story with the rest of the house.

Dr. Telkes admitted that as the first of its kind, her system could not be expected to work perfectly. "Who can expect the first of its kind to be 100 percent effective?" she asked. Perhaps further, more rigorous testing would have allowed her to dispel the cloud of doubt that surrounded the use of Glauber's salts after the Dover experiment. In retrospect, total reliance on the sun for house heating in a cold, cloudy climate like that of Massachusetts was an unrealistic goal, and a backup heating system should have been installed at the outset.

A Solar-Heated Schoolhouse

Compared with Massachusetts and Colorado, the winters in Tucson, Arizona, are like spring; the average daytime temperature only drops to the mid-50's in January. A solar heating system there could be expected to carry a much higher percentage of the heating load—especially if heating at night were not required. Such was the case in the first solar-heated public building, Rose Elementary School, which was designed by Arthur Brown and built in 1948. When he was first commissioned to work on this project, Brown knew that his main challenge would be to provide a solar heating system without running up a horrendous bill that would antagonize the taxpayers. His solution was unique and effective.

There was no need for heat storage (a major expenditure in most systems) since classes went from 9 a.m. until 3 p.m.—the warmest, sunniest part of the day. Equally important in cutting costs, the roof itself served as the solar collector, and the conventional hot-air system distributed the solar heat. The roof consisted of aluminum troughs covered with an aluminum sheet to form parallel air channels. A fan circulated the room air through a duct and into the channels. The air warmed up

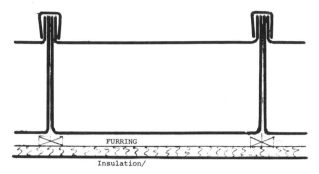

Left: Cross-section of the collector used in the Rose School. Aluminum troughs doubled as roof supports and ducts for the solar-heated air.

Below: Rose Elementary School in Tucson, Arizona. Arthur Brown designed this first solar-heated school building in 1948.

by 10 or 15°F and a second fan sent this warm air to the classrooms. This solar heating system provided 86 percent of the school's heat. Even higher temperatures could have been attained if Brown had covered the roof with one or two layers of glass, but his budget was severely limited.

In the summer, Brown's design helped to keep the school buildings cool with a minimum of artificial air conditioning. A longtime resident of Arizona, he remembered that many old-time residents put a second roof over their ordinary roof to help keep out the sun's heat in summer. He applied this concept to the school building by relying on the double aluminum channels to help shield the classrooms from the sun's burning rays. The ducts that normally circulated solar-heated air to the rooms during winter could be opened manually in summer to let the fans blow the warm classroom air outside. Brown also had the roof built so that it extended beyond the south wall of each building. Thus it shaded the walkway beside the building and the south walls of the classrooms inside.

For ten years Brown's system kept the Rose School warm in winter and cool in May and September—the two hottest months of the school year. But when the time came to expand the complex, the local school district decided to replace the system with a gas furnace. Some teachers had complained that the fans made too much noise, and others did not like the fact that the ducts had to be opened by hand. No one thought of saving energy, Brown noted regretfully. "I did these things at a time when gas was so cheap," he remarked, "that people didn't have an interest in solar heating." But his system had demonstrated a workable approach to solar heating for a building located in a mild climate and occupied only during the daytime. Its simplicity kept construction and operating costs at a minimum. Since the collector also served as the roof, and the only extra expense was that of keeping the fans running, the system began paying dividends from the moment it was installed.

The Fourth M.I.T. House

During the 1950's, solar architects and engineers built a second generation of buildings heated by "active" solar heating systems. Most of these projects were genuine dwellings or office buildings designed to demonstrate that comfort, convenience and aesthetics need not be sacrificed by the use of solar collectors for space heating. George Löf used another solar hot-air system, similar to the one he had developed in Boulder, to heat his newly built, ranch-style home in Denver, Colorado. In Albuquerque, New Mexico, the engineering firm of Bridgers and Paxton built the first solar-heated office building in 1956. This system could also be used to cool the building in summer.

In 1958, the M.I.T. solar research team erected a fourth solar house—this one a full-scale dwelling built from scratch in Lexington, Massachusetts. Once again they used solar-heated water to warm the house. But this time only about half of the house's heat was supplied by the sun. The M.I.T. staff conducted a very sophisticated economic analysis of this system. On the basis of the low cost of fossil fuels at the time, they calculated that the cost of the solar heating system would have to be slashed by 80 percent if it was to pay for itself in ten years. According to Hottel, "With the price of oil as it was then, solar would [have been] economically interesting only if we could have gotten the collectors for a couple of dollars a square foot." This was impossible—given the high cost of aluminum, copper and glass. George Löf agreed with Hottel's assessment of solar economics in the 1950's.

But despite these pessimistic views, the years of research had shown that solar house-heating in a cold climate was technically feasible. The only stumbling block was cost. Perhaps that obstacle could have been overcome by taking the inexpensive water wall of the second M.I.T. experiment and developing this approach further. Or perhaps the research team should have learned from the success of their third house and incorporated large south windows into the Lexington house to supplement the heat trapped by its collectors. But the research team failed to consider these options, and in 1962 they terminated more than two decades of hopeful research.

The Engbretson family, residents of M.I.T.'s fourth solar house, roasting hot dogs in a solar oven on their front lawn.

V
The Modern Era

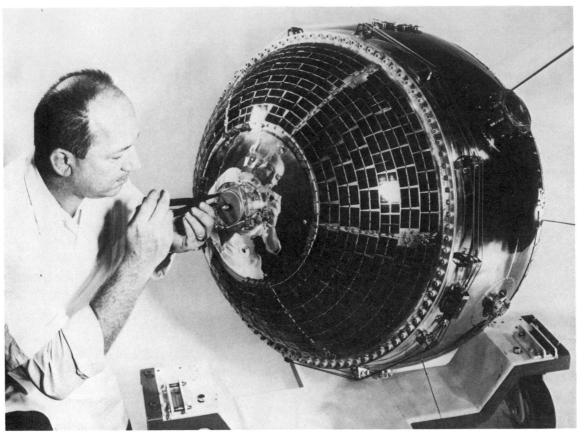

Almost every American space satellite, including the Explorer-B shown here, used silicon solar cells for electric power.

Chapter 17
Postwar Energy Perspectives

For almost three decades after the end of World War II, the United States had few problems of energy supply. Its industry, commerce and homes all had ready access to oil and gas from both domestic and foreign sources. Most of the oil was close to the surface, easy to tap, and therefore economical to extract. Foreign governments sold their oil to American companies at extremely low prices. The U.S. government also helped to keep oil prices low and profits high. Depletion allowances permitted oil companies to write off a portion of their income taxes against the cost of discovering and drilling for oil in the United States. Royalty payments on foreign oil could also be deducted. These and other government subsidies helped to keep the price of oil below $3 a barrel during the late 1950's and throughout the 1960's. Natural gas prices were low, too—below $1 per thousand cubic feet during this same period. Often obtained as a by-product of oil exploration, gas enjoyed the same tax advantages as oil. The cost of natural gas was also kept down by government regulation of gas prices and by improved pipeline networks that made supplies much more accessible.

The falling prices of fossil fuels during this period also reflected the growing availability of supply. Crude oil reserves rose steadily between 1952 and 1964, as did natural gas reserves. Corporate spokesmen assured the public that this rosy situation would continue almost indefinitely. As one gas company executive stated in 1954, "The industry discovers more gas every day than is consumed every day in the United States. I don't think in our lifetime we will see the depletion of our product."

With fuel apparently so abundant and cheap, electric companies began to expand to meet growing postwar demands. Liberal government policies made it easy to procure the needed capital to build larger and more efficient power plants. Soon after World War II the electric rates plunged as consumption grew and power plant efficiency increased. The electric utilities encouraged greater consumption because the costs of building new plants and installing electric lines could be recovered more quickly if their customers used more electricity. "Once you had the lines in, you hoped that people would use as much electricity as possible," an executive for one electric company remarked. "You wanted to get as much return on your investment as you could. The gas companies had a similar objective. As one employee explained, "If you sell more you make more."

Both gas and electric utilities promoted consumption through advertising campaigns and preferential rate structures. "The Gold Medallion Program," a national promotional campaign conceived by General Electric and later taken over by regional electric utilities, urged Americans to buy more electric appliances—and thereby use more electricity. Lower rates for increased use of electricity also stimulated consumption. During the 1960's many U.S. families paid about 4¢ per kilowatt-hour if they used 100 kilowatts or less per month, but the rate fell to 2¢ if more than 750 kilowatt-hours were used. In some areas, people paid less than 2¢ per kilowatt-hour. The gas companies had similar rate structures that made it cheaper to use more gas. They also launched their own pro-consumption campaign, which used the "Blue Flame" as its slogan and symbol.

The energy companies' publicity and the enticement of lower prices worked. A desire to "Live Better Electrically" led families to opt for homes with electric house heating, water heating, ovens and many other appliances. A growing

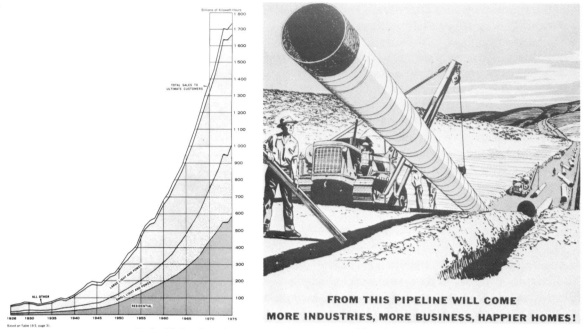

Left: U.S. electric consumption increased by almost a factor of five during the 1950's and 1960's.

Right: A 1955 advertisement for the Southern California Gas Company. Improved pipeline networks made natural gas available in millions of American homes during the same two decades.

affluence that allowed people to indulge their appetite for new electric appliances, combined with the postwar baby boom, helped increase electricity generation by over 500 percent between 1945 and 1968. Natural gas consumption also zoomed upwards as gas heating and conveniences such as clothes dryers became more popular. Natural gas production more than doubled—from 6 to 16 trillion cubic feet—between 1950 and 1965. U.S. fuel consumption as a whole more than doubled between 1945 and 1970.

A Note of Caution

The frenetic pace at which America was gobbling up its energy resources alarmed only a few farsighted individuals. Eric Hodgins, editor of *Fortune*, called the careless burning of coal, oil and gas a terrible state of affairs, enough to "horrify even the most complaisant in the world of finance." Writing in 1953, he warned that "we live on a capital dissipation basis. We can keep this up perhaps for another 25 years before we begin to find ourselves in deepening trouble." But such warnings were generally treated with derision or merely ignored. Those predicting energy shortages were labeled pessimists. "Not many in industry wanted to hear such talk," commented Charles A. Scarlott, then editor of the Westinghouse Company's technical publications. "They were making too much money on energy sales."

A few scientists and engineers took the same dim view as Hodgins and sought an alternative to the fuel crisis they saw was inevitable. In 1955, they founded the Association for Applied Solar Energy Research and held the World Symposium on Applied Solar Energy in Phoenix, Arizona. Delegates from all over the world attended, presenting research papers and exhibiting solar devices. Israel displayed its commercial solar water heaters, and representatives from Australia and Japan discussed their nations' increasing use of the sun. To many, the Symposium represented the dawn of a new solar age. But the careless confidence of energy-rich America squelched the hope in that country. Solar energy received virtually no support in the ensuing years, and by 1963 the Association found itself bankrupt. "They couldn't even pay my final salary," noted Scarlott, then editor of its publications.

Whereas the governments of Israel, Australia and Japan deliberately aided the solar industry, the U.S. Congress and the White House were sitting on the sidelines while the hopes of a prescient few floundered. True, as early as 1952 the President's Materials Commission appointed by Harry S. Truman came out with a report, "Resources for Freedom," predicting that America and its allies would be short of fossil fuels by 1975. This report urged that solar energy be developed as a replacement. "Efforts made to date to harness solar energy are infinitesimal," the commission chided, despite the fact that "the United States could make an immense contribution to the welfare of the free world" by exploiting this inexhaustible supply. The commission predicted that, given the will to go solar, there could be 13 million solar-heated homes in the nation by the mid-1970's. However, Washington did not heed this advice in the years that followed. Federally funded scientific research had a multi-billion dollar budget, but the amount allocated to solar energy projects was only a tiny fraction of one percent.

The Nuclear Genie

The U.S. government admitted the very real possibility of an energy crisis. But it chose a different solution—the atom. Beginning in the early 1950's with the Eisenhower Administration, nuclear power was seen as the energy source of the future. Energy equipment manufacturers and utility companies took their cue from Washington and jumped into nuclear development. According to a 1954 article in *U.S. News and World Report*, the huge corporations involved in the new nuclear industry were "backed by great capital resources and staffed with scientists," and could take advantage of "the knowledge already acquired by 14 years and 10 billion dollars worth of federal research."

Nuclear energy appeared to offer tremendous power in exchange for relatively small amounts of fuel and labor. Scientists and the general public alike shared this belief, and many people around the globe were infected by America's enthusiasm. At an international conference on nuclear energy held in 1954 in Oslo, Norway, Alvin M. Weinberg, then director of Oak Ridge National Laboratory in Tennessee, reported that Europeans seemed even more excited about nuclear prospects than their American colleagues. He found them quite concerned about their import-dependent situation. Indigenous coal supplies were inadequate to meet projected energy needs over the next two decades—meaning further reliance on imported oil.

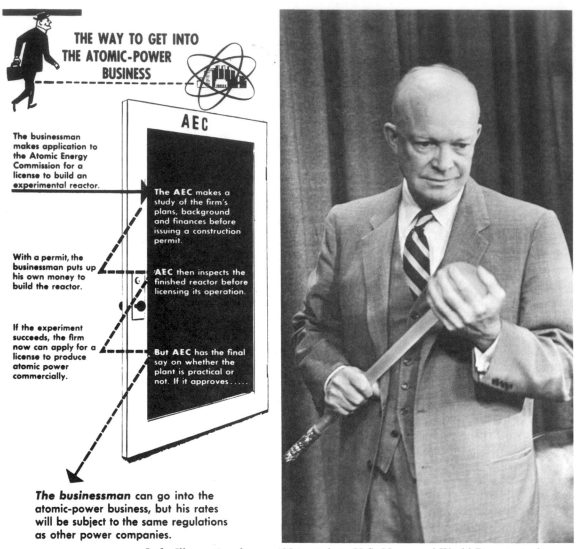

THE WAY TO GET INTO
THE ATOMIC-POWER
BUSINESS

AEC

The businessman
makes application to
the Atomic Energy
Commission for a
license to build an
experimental reactor.

The AEC makes a
study of the firm's
plans, background
and finances before
issuing a construction
permit.

With a permit, the
businessman puts up
his own money to
build the reactor.

AEC then inspects the
finished reactor before
licensing its operation.

If the experiment
succeeds, the firm
now can apply for a
license to produce
atomic power
commercially.

But AEC has the final
say on whether the
plant is practical or
not. If it approves.....

The businessman can go into the
atomic-power business, but his rates
will be subject to the same regulations
as other power companies.

Left: Illustration from a 1954 article in U.S. News and World Report. *A close
partnership between government and industry made the dream of nuclear power
a reality.*

*Right: In 1954, President Eisenhower inaugurated the construction of America's
first nuclear power plant with a wave of this radioactive wand.*

Two years later, with the Suez crisis threatening to strangle Europe's oil supply line,
leaders in England and on the Continent fervently embraced nuclear energy as their
savior. As *Business Week* reported, they were determined that their "industrial
machine and livelihood must no longer be at the mercy of Middle Eastern politics."
Even such countries as India and, ironically, Japan took up the nuclear banner—
although solar water heaters were already helping thousands of Japanese save fuel
at the time.

How the atom is putting new shapes on the horizon

Look *magazine advertise-ment for the electric utilities, 1956. Nuclear energy prom-ised the utilities a centralized power source with little labor involvement and low fuel cost.*

Meanwhile America's nuclear program was rapidly advancing. It had the blessings of President Dwight D. Eisenhower, who preferred his own "Atoms for Peace" plan to the recommendations of Truman's Materials Commission. The media coverage of Eisenhower's ceremonial inauguration of America's first com-mercial nuclear power plant illustrated the nation's heady romance with the atom. In 1954, *Life* magazine described the event:

> With a wave of a radioactive wand, President Eisenhower transformed the bright hope for atomic power peaceably used into a solid certainty. . . . Standing in the studio of Denver's T.V. station KOA, the President slowly lowered the head of the wand over a fission counter. When the counter's needle swung across the dial, it electrically set in motion, 1,300 miles away at the Shippingport, Pennsylvania, plant site, an automatically controlled power shovel which scooped up the first symbolic shovelful of earth.

Two years later, England opened its first commercial nuclear power plant with similar fanfare.

The entire U.S. political spectrum backed nuclear power as the dream solution to future energy problems. Government officials, industry spokesmen and nuclear scientists claimed that atomic power was a "clean, safe" energy source. It would eliminate the air pollution inherent in oil-fired and coal-fed power plants. The only controversy was over who should own and operate the reactors—private enterprise or the government.

Only a few individuals could see that the Aladdin's lamp of nuclear power was flawed, and fewer dared to say so. In 1956, Erich A. Farber and J.C. Reed of the College of Engineering at the University of Florida stressed that nuclear is limited "both by uranium ore supply and by the inherent hazards of radiation." George W. Russler, Chief Staff Engineer at the Minneapolis-Honeywell Research Center, analyzed the push toward nuclear power as an unfortunate choice. In an article, "Nuclear or Solar Energy," published in 1959, he commented on the waste disposal dilemma:

> If one projects the problem into the future when all the world's conventional power plants, multiplied by a factor of 23 or more, are replaced by atomic plants, then the enormity of the problem of waste disposal becomes apparent. Perhaps, on this scale, the problem may not be solvable.

Another problem was the enormous amount of energy needed to enrich uranium. According to experts cited by Russler in 1959, 10 percent of America's electrical energy was being used to produce the nation's output of enriched uranium—most of which was used to manufacture nuclear warheads. He questioned whether there would be any net gain in electrical production if this enriched uranium were instead used to produce electricity.

Russler contrasted the problems of nuclear energy with the attractiveness of solar energy:

> Solar energy is the one major source of energy which would not require several decades of development before large energy contributions could be obtained. Its use does not involve such serious problems as the control of a critical mass, or disposal of dangerous waste products, or operating health hazards. It does not require multi-billion dollar installations, nor huge concentrations of basic materials, nor elaborate controls. Sufficient engineering know-how, as well as simple processes, are already sufficiently available to make a major start at its utilization. . . . The only elements lacking are an appreciation of the urgency of the energy situation and a determination to get started.

Russler suggested that solar energy could make a large contribution toward heating buildings. In America, this task consumed larger quantities of fossil fuel than transportation, industrial processes, or electrical power generation. He pointed out that the low-temperature heat needed in homes and office buildings "ideally matches the low-grade heat derived from the simplest and most efficient solar energy collectors." Hence, this was the perfect way to start putting solar energy to widespread use.

The Discovery of Solar Cells

At about the same time that the first commercial nuclear power plants were being built, vastly improved photovoltaic cells were developed at Bell Telephone Labs in Murray Hill, New Jersey. These "solar cells" could transform solar energy directly

*Edmund Becquerel,
the French scientist
who discovered the
photovoltaic effect
in 1839.*

into electricity. Toward the end of his career, Augustin Mouchot had experimented with solar generation of electricity, but he was using the action of solar heat on dissimilar metals to produce an electric current. In photovoltaic cells, it is the light rather than the heat of the sun that generates electricity.

In 1839 Edmund Becquerel, a French experimental physicist, discovered that sunlight could produce electricity. Almost 50 years later Charles Fritts, an American inventor, made the first solar cells. These thin wafers—each about the size of a quarter—were made from selenium (an element derived from copper ore) covered with a transparent gold film. When sunlight struck the cells, a current was generated that was "continuous, constant, and of considerable electromotive force," according to Fritts. He believed that at least 50 percent of the light hitting the surface of these cells could be converted into electrical energy. But such a grandiose forecast proved unrealistic. He did not realize that less than one percent of the light energy hitting the selenium was actually converted into electricity, and that the mechanism for capturing this electricity was no more than 50 percent effective.

During the next few decades, few people took any interest in trying to upgrade the performance of solar cells. Classical physics of the late nineteenth century could not explain the photovoltaic effect, and many scientists and engineers did not give it any credence. Only after the bold new theories of quantum mechanics and relativity won general acceptance in the early twentieth century did work in solar cells begin again. Scientists now pictured an electrical current as an orderly movement of electrons which could be set into motion by direct interaction with particles of light called photons.

With these firm theoretical underpinnings, scientists began to reexamine the photovoltaic effect, and rediscovered the selenium solar cell in the early 1930's. Aside from minor design changes, this cell was almost an exact replica of the one developed by Fritts. Its reappearance renewed the dream of producing electricity commercially without fuel. However, scientists were hindered by the same limitations that Fritts had encountered earlier: the amount of electricity produced was miniscule. The rediscovery of the solar cell did lead to useful light-sensitive devices such as the photometers used in photography. But, for more than two decades the best selenium cells could convert less than one percent of all incoming sunlight into electricity—hardly enough to justify their use as a power source.

In 1954 researchers at Bell Telephone Laboratories made an accidental discovery that revolutionized solar cell technology. They were searching for a dependable alternative power source for telephone systems in rural areas. Darryl Chapin, leader of the project, thought that improved selenium cells would be the ideal solution. But efforts to develop a more efficient selenium cell failed. Meanwhile Calvin Fuller, a Bell scientist working in another department, had been studying silicon—one of the two major elements in common sand. Fuller was exploring

Bell Telephone scientists Gordon Pearson, Darryl Chapin and Calvin Fuller measure the response of a silicon solar cell to light.

silicon's usefulness in making a rectifier, a device that changes alternating current into direct current. Fuller discovered that he could increase the efficiency of the silicon rectifier by adding certain impurities. The director of Fuller's project, Gordon Pearson, happened to expose the modified silicon rectifier to light. To his surprise, a significant electric current was generated.

Pearson knew the problems Chapin had been having with the selenium cells. So he brought the discovery of silicon's light sensitivity to Chapin's attention, and soon Fuller and Chapin were busy refining a silicon solar cell. Their first design could convert 4 percent of all the incoming sunlight into electricity—five times more than the best selenium cell. Not content with this conversion ratio, they continued to work on the silicon cell for several months. By May they had produced a solar cell with an efficiency of 6 percent. News media carried pictures of Chapin, Fuller and Pearson powering a transistor radio with solar energy. The response of people around the world was tremendous. *Business Week* ran a futuristic article about solar-powered fans, lawnmowers, and even a solar convertible that was steered automatically so that "all the riders could sit comfortably in the back seat and perhaps watch solar-powered TV." Others rhapsodized about acres of solar cells supplying the world with cheap, nonpolluting energy.

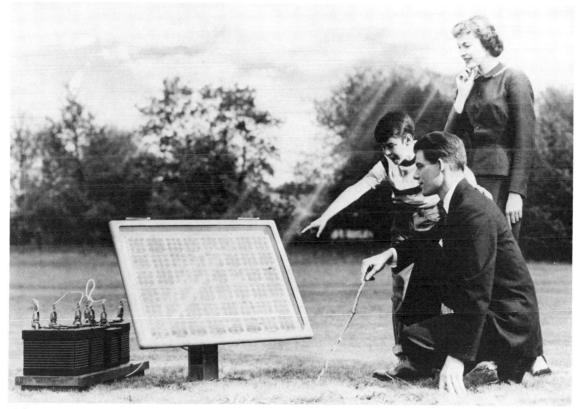

Advertisement from Look *magazine, 1956. "Bell System Solar Battery Converts Sun's Rays into Electricity!" read the copy.*

Right: This array of silicon solar cells powered a rural telephone system in Americus, Georgia.

Below: A telephone lineman installing the solar cells.

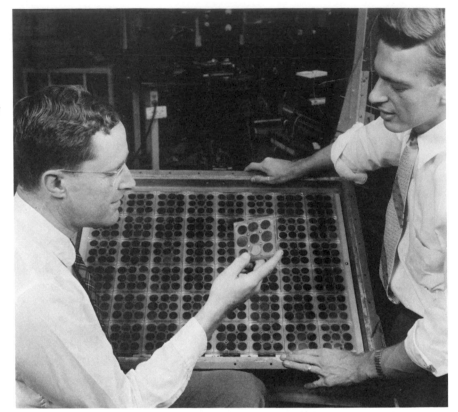

Fuller and Chapin took a more sober view of the silicon cell, even after reaching a conversion ratio as high as 15 percent. "We tried to avoid making too much claim for it," said Chapin, "because we knew it was in the laboratory stage, it was an expensive process, and there was much to be done before we could speak of lots of power." The high-purity silicon required for efficient cells cost $80 a pound at the time. Furthermore, each razor-thin wafer had to be individually sliced from its component core, which hiked manufacturing costs even higher. And the power output per cell was so small that providing large amounts of electricity would be prohibitively expensive for most commercial purposes.

Solar Cells at Work

Silicon solar cells were first used as the power source for a telephone relay system in an isolated rural area of Georgia. The setup included battery storage for a nighttime power supply and worked without problems. However, Chapin's concern about the cost of solar-generated electricity proved to be well founded. An economic analysis of the solar-powered telephone system showed that it was not competitive with a system powered by conventional electricity.

Just as solar cells were about to be consigned to the curiosity heap, the space race came along. Satellites needed a long-term, autonomous power source that was compact and lightweight. Conventional fuel systems or batteries large enough to supply the energy required were too cumbersome. The National Aeronautics and Space Administration saw solar cells as the perfect answer. They did not have to be connected to a storage system because the sun shines 24 hours a day in outer space.

Women assembling solar cells for the U.S. space program.

Silicon solar cells were a durable, compact, lightweight source of electric power for this Explorer space satellite.

Hence, solar cells were easily the lightest power source available, and they proved cost-effective for space applications. The U.S. space program created an entire solar cell industry. Starting in the late 1950's, solar cells powered all American space satellites from Vanguard to Skylab.

Back on earth, however, the terrestrial use of solar cells did not receive any support whatever. Fossil energy consumption continued to break records, even though there were increasing signs that severe fuel shortages loomed in the near future. Domestic oil production in the United States increased by 43 percent between 1953 and 1969, but the number of new oil discoveries fell by 43 percent during that period. America's reliance on imported oil also grew, rising from 14 percent in 1954 to 22 percent in 1965. Two years later, the size of America's crude oil reserves declined for the first time in the nation's history.

Despite such portents the government never funded research to develop better and cheaper solar cells for ordinary commercial use. Why it showed such a lack of interest was a mystery to some. As early as the mid-fifties, the *New York Times* suggested that the U.S. government "ought to transfer some of [its] interest in atomic power to solar." But Washington's attitude mirrored that of a nation hypnotized by seemingly limitless supplies of cheap fossil fuel, and by the almost magic aura surrounding nuclear energy. There was no solar lobby to counter the already powerful nuclear juggernaut. Consequently, solar cells and other solar technologies received very little support during the quarter century after World War II. The concept of sun power remained mostly within the realm of science fiction.

More than two dozen Miromit collectors can be seen on the roofs of these dwellings at Benai Hadarom, an Israeli kibbutz near Tel Aviv.

Chapter 18
Worldwide Solar Water Heating

Not all areas of the world enjoyed the cheap, abundant energy supplies that allowed Europe and America to forget the sun in the years after World War II. Many countries did not have easy access to the river of oil issuing from the Persian Gulf states and other oil-rich nations. Solar water heating industries were established in some of these fuel-short regions—especially in areas with ample sunshine. Business flourished for manufacturers in Israel and Japan during the 1950's and 1960's. Solar water heaters were also successful in parts of South Africa and Australia.

The solar industry had a humble beginning in Israel—a mother needed hot water to bathe her infant son. The mother was Rina Yissar and the year was 1940—a time when extreme scarcities of fuel oil plagued Palestine. Most people were forced to accept cold baths as yet another sacrifice to help found the Jewish state. But Rina, whom her son Gonen later described as a woman who "lacked formal technical education but had excellent common sense," refused to resign herself to this hardship. Instead, like many rural folk in the United States years before, she took an old tank, painted it black, filled it with water, and left it out in the sun. After a few hours, she had enough hot water to give her baby a warm bath.

This simple demonstration of solar water heating fascinated Rina's husband Levi, a civil engineer. At her urging, he began to look into the matter, but World War II and then the bitter battles of 1948 forced him to put off his research. When a shaky peace finally arrived, Levi Yissar resolved to devote all of his efforts toward harnessing the sun. He scanned the technical literature and found several professional accounts of solar water heating in California and Florida, as well as the study of solar collectors published by Hottel and Woertz in 1941. He also attended an international conference on solar energy held at M.I.T. in 1950.

Returning home, Yissar combined what he had learned about solar collection and heat storage with several of his own innovations and built a prototype solar water heater. Essentially, his collector and storage tank resembled William Bailey's Sun Coil design; like those in California and Florida, his system relied on thermosyphoning to circulate water between the two. Yissar used the same type of metal—steel—for the tubing in both the collector and the storage tank. He wanted to avoid the corrosion problems that had caused the tanks to rupture in Miami. And to increase the collection efficiency he inserted a dehumidifier, which reduced the amount of moisture that accumulated inside the collector box in the extremely muggy summer weather of the Israeli coastal plain where most of the people lived.

Yissar's first efforts at promoting solar heating encountered stiff professional resistance. Despite his claim that solar energy could help conserve precious fuel reserves (all fuel in Israel was then imported), his colleagues remained skeptical. "Everyone laughed at me," he recalled. "No one believed that the sun could produce water hot enough for general household use." But Yissar ignored his critics and soon built his first heater, which met all the hot water requirements of the family living in the house where it was installed. Soon 25 more heaters were set up in his home town of Holon, a suburb of Tel Aviv. The business community was impressed—the heater was built to last ten to fifteen years, and would pay for itself in only two. The only alternative, electric water heating, was extremely expensive in fuel-short Israel. In 1953 Yissar obtained sufficient capital to establish the Ner-Yah Company, Israel's first manufacturer of solar water heaters. One of his first customers was David Ben Gurion, the founding father of Israel, who had a solar

Top: Levi Yissar (left), the civil engineer who introduced the solar water heater in Israel. One of the first commercial units (right) he installed in his hometown of Holon, a suburb of Tel Aviv.

Above: The solar water heater marketed in Israel by the Ner-Yah Company; Yissar is shown standing behind it. Cross section of the system (right).

heater installed in his home. Within a year Yissar's company sold 1,600 solar water heaters.

Miromit Enters the Field

Eyeing Yissar's success, several formerly doubting colleagues decided to establish their own solar companies. Yissar's son Gonen claimed that "some of them got into the market with their own 'inventions' by violating my father's rights on his patents." As in Florida, many of these new companies built substandard equipment in their scramble to gain a competitive advantage. This trend worried the Israeli

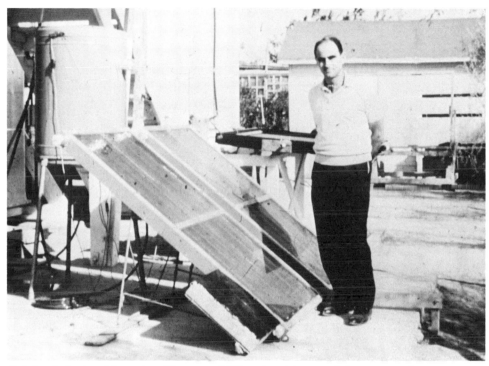

Rainier Sobotka, Managing Director of Miromit, with one of the company's high-quality solar heaters.

government because many of the heaters were being installed in government buildings—mostly housing projects for newly-arrived immigrants. The government finally intervened and worked out a plan whereby Miromit, Israel's largest metal fabricator, received exclusive rights to Yissar's patent on the condition that it maintain high production standards. The government also awarded Miromit a license to use a special coating called a "selective surface" on the absorber plate. Invented by another Israeli, Harry Tabor, this coating inhibited thermal radiation from the absorber and thereby cut heat losses from the collector by 30 percent.

Miromit also introduced several improvements of its own. The serpentine coil arrangement of the collector tubing was replaced by parallel piping. This change insured better heat transfer from the absorber plate to the water and made it easier—and therefore cheaper—to flush out the pipes when they became clogged with mineral deposits. Israel's hard water was also the primary reason for not soldering the collector pipes to the absorber plate. Instead, the tubing fit tightly into grooves in the absorber. These pipes could be readily scrapped when they became too badly blocked by minerals, but the absorber plate could be reused. Miromit also made a collector box with a tighter seal to keep out the rain and dust.

Miromit's high-quality product helped upgrade the heaters marketed by other Israeli companies. According to Harry Tabor, "Keeping up with Miromit" became the song that many others in the industry sang. In fact, their competitors conformed so slavishly to every new Miromit development that when the color of the collector boxes was changed for purely aesthetic reasons, all the other companies followed

Roger N. Morse,
who introduced solar
water heating in
Australia.

suit. "Presumably," mused Tabor, "they assumed that there was some scientific reason for the change!"

Nearly 50,000 solar water heaters were sold in Israel from 1957 to 1967. Miromit also ran a booming export business. According to Managing Director Rainier Sobotka, the company shipped heaters to "the Canary Islands, Reunion, and Honduras, to Trinidad and the St. Vincent Islands in the Caribbean, to Manila and Singapore, to Iran, Turkey and Chile—all in all, to about 60 different countries" by 1961.

Ironically, victory in the Six Day War of June, 1967, had a devastating effect on Israel's solar water heater industry. The nation captured large oil fields on the Sinai Peninsula—ending decades of fuel scarcity. When this oil began to flow toward Jerusalem, people stopped buying solar water heaters. Once again the siren call of fossil fuels led people away from the sun.

The Australian Government Lends a Hand

Australia's government played an active role in getting solar water heating started there. Roger Morse, an engineer with the Commonwealth Scientific and Industrial Research Organization (CSIRO), began to develop a government program in 1952. He and his colleagues developed a solar collector very similar to the model used at M.I.T.—basing their design on the seminal paper written by Hottel and Woertz. They also encouraged private industry to enter the field, and a number of firms soon began to **manufacture** and sell solar water heaters. Government

This group of solar heaters provided hot water for the Don Hotel in Darwin, Australia.

advisors visited the factories and maintained a close rapport with the fledgling companies. But many of their early systems functioned poorly or did not work at all, because plumbers had difficulty installing these solar water heaters. To rectify this problem, Morse wrote an installation manual that became the industry's standard guide.

Aside from providing technical advice, the government also developed a market for solar heaters. In the mid-fifties, a government study recommended that all state buildings in tropical areas be equipped with solar water heaters. These areas had the highest electric rates in Australia and usually enjoyed the most sunshine. Morse called this government attitude "the catalyst which gave encouragement to the industry." The resulting demand helped two major solar firms to prosper—Solahart and Beasley Industries.

After marketing a thermosyphoning system for several years, Solahart began to use pumps to circulate water from the collector to the storage tank. But the company soon found the pumped system unreliable. Keith Jenkins, who worked with Solahart, commented: "When the pump goes bad . . . the whole system breaks down. And like all things that break down, they break down at the most inconvenient times." Also, people turned the pumps off when they went away on vacation. With no water flowing through them, the collectors got extremely hot, damaging themselves. It didn't take long for the company to return to the thermosyphoning system, which was much less troublesome.

Solahart also chose a completely new integral collector-tank configuration to facilitate easy installation on pitched roofs. The storage tank was mounted along the upper edge of the collector. Not only did this new arrangement eliminate the need for extensive piping; it also removed the necessity of installing a heavy storage tank in the attic. Soon other firms adopted this modular system. Solahart, along with Beasley and other firms, slowly built up their market—primarily through word-of-mouth promotion by satisfied customers. Between 1958 and 1973, about 40,000 solar collectors were sold throughout Australia.

An early Solahart water heater installed in the Cocos Islands, circa 1960.

Problems in South Africa

Solar water heating met quite a different fate in South Africa. Lewis Rome, an English-trained engineer living in Johannesburg, founded the Economic Solar Water Heater Company in 1954. Rome had learned about solar heating from Austin Whillier, a fellow South African who had studied under Hoyt Hottel at M.I.T. Thus it is not very surprising that his collector also resembled the M.I.T. model.

Most of Rome's customers lived in rural areas, primarily in the northern Cape region and in Southwest Africa—then a mandate of South Africa. The high cost of shipping fuels to such remote areas made solar water heating economically attractive there. On the other hand, coal prices and electric rates in the country's urban centers were then among the lowest in the world. Even here, however, the payback time on a solar water heater was only seven years. Apparently that seemed too long a time for many city residents, for Rome's sales were poor in South Africa's cities.

Some progressive citizens felt that cost should not hinder South Africa's full-scale use of solar water heaters. They urged the government to push solar energy to help alleviate pollution caused by coal burning. "For every solar heater installed where coal had been previously used," declared one engineer, "smog must become proportionately less." But the government did not act. On the contrary, a bureaucratic measure killed solar energy's economic appeal even in rural areas. In 1961 the government doubled the rail rates for all manufactured goods and reclassified solar

Advertisement for Economic Solar Heaters, South Africa's first solar industry.

heaters so that they were subject to the highest transportation tariffs. The cost of shipping them from Johannesburg to outlying areas now outran production costs—boosting their list price beyond the reach of most people. The loss of this market forced Rome out of business that year.

The Japanese Industry

Like the Romans, the Japanese have always loved their hot baths—especially farmers coming home covered with mud after spending long hours in the rice fields

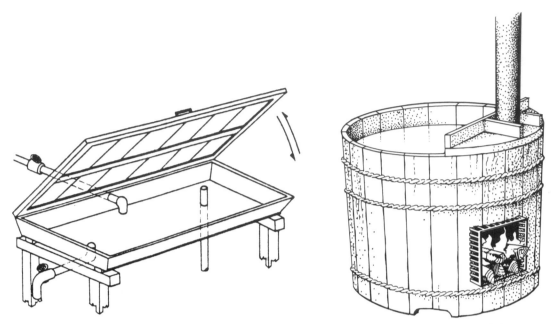

Left: Japan's first solar water heater, invented by Sukeo Yamamoto in 1947.

Right: Artist's rendering of a wooden Japanese bath tub. Large amounts of scarce fuel were used to keep the bath water warm.

during the hot, muggy summers. They usually bathed in large wooden tubs heated by a fire underneath. Wood was abundant in the mountains, but in the low-lying districts people had only rice straw for fuel. During the Depression years of the 1930's, some farmers in fuel-short regions began to use the sun for heating their bath water. However, none of these early attempts spread beyond the local level.

On a trip to the countryside during the 1940's, Sukeo Yamamoto saw one of these primitive heaters—a large bathtub filled with water whose top was covered by a sheet of glass. When set out in the sun early in the morning, it produced water hot enough for bathing by about two in the afternoon. Yamamoto was taken with the simplicity of this device, and two years after the war's end he designed Japan's first commercial solar water heater. It consisted of a rectangular wooden basin measuring 6 feet long, 3 feet wide and 6 inches deep. Glass covered the top and a thin sheet of blackened metal lined the interior. This heater was usually mounted horizontally near the bathtub, with an unobstructed exposure to the sun. In the morning a faucet was opened to fill the basin with water. At night the sun-heated water was emptied into the bathtub via another faucet. From late spring through the early fall, the 53-gallon heater could provide bath water up to 140°F by late afternoon—the customary bath time. From November to March, when the low-lying sun cast its oblique rays on the heater, the water temperature reached only 70 or 80°F. Auxiliary heat was then used to raise the water another 30°F or so for a good steaming bath.

The Kaneko-Kogyosho Company began mass-producing Yamamoto's heater in 1948. Farmers in the valley rice regions found it ideal—it was simple to operate, worked fairly well, and annually saved each family about 1½ tons of rice straw,

The vinyl plastic "air-mattress" solar water heater, introduced in the late 1950's. A plastic canopy supported by the wire mesh improved its performance.

which could then be used as cattle fodder or fertilizer. The heater cost only $20, and farmer's associations and government authorities helped the farmers finance the heaters with direct subsidies and low-interest loans. The solar heater quickly caught on and became popular in farming villages throughout the country. By 1955, over 20,000 had been installed in Japan.

Yamamoto's invention did have its drawbacks. Many found that the loose-fitting glass cover allowed too much dust and dirt to contaminate the water in the basin, and that algae often grew in the stagnant warm water. Competitors anxious to enter this growing market put out a completely sealed water heater that avoided these problems. Made of soft vinyl plastic, this heater was the same size as Yamamoto's but resembled an inflatable air mattress. Two models were available: one with clear plastic on top and black plastic on the bottom, and the other made completely out of black plastic. Both styles produced hot water about as well as Yamamoto's original heater. In winter a clear plastic canopy could also be put over either version to help retain the solar heat. Some left the canopy up year-round.

The very low price of the soft vinyl heater—ranging from $6 to $10—made it accessible to almost everybody. Urban department stores found them convenient to market because they could be folded up and sold in small cardboard boxes. Consumers also appreciated their easy installation. The vinyl heater sat on a flat wooden base built by the customer and could be readily connected to the household

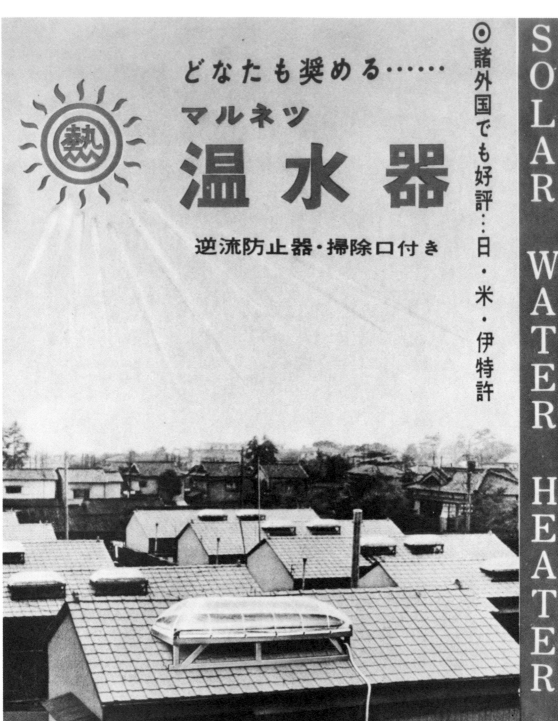

This 1960 advertisement urged the Japanese to "shower with solar-heated water."

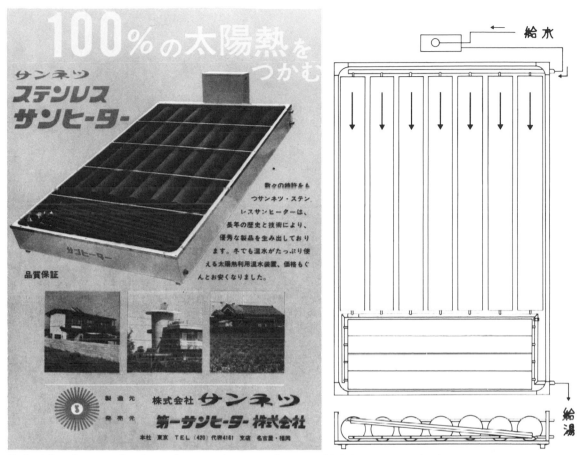

Left: Advertisement for the tank-type solar water heater, which became popular in Japan during the 1960's.

Right: Flow diagram for the tank-type heater. Cold water enters the tanks from a reservoir and warm water is drawn off at the outlet.

water supply by plastic inlet and outlet hoses. True, the plastic lasted only two years or so. But as Professor Ichimatsu Tanashita, a leading Japanese authority on solar energy, pointed out, the amount of fuel and labor that was saved during those two years far outweighed the cost of the heater. The Japanese public took an immediate liking to this solar water heater, buying 20,000 in 1958, the first year they were sold, 70,000 in 1959, and 170,000 in 1960.

More and more companies were eager to take advantage of the growing solar demand among Japan's farmers and the burgeoning middle class. Newcomers to the field looked for ways to enter the market, which until the late 1950's had been dominated by one firm. In 1960, they came up with an improved solar water heater, one that was more solidly built so that it would last longer than the soft vinyl heater. It could also be placed on an incline toward the south to receive the sun's rays more directly in winter. Without knowing about early American solar water heaters, the Japanese designed a device bearing a very close resemblance to the old Climax

Professor Ichimatsu Tanashita, an early Japanese advocate of solar energy. He tried to introduce the flat-plate collector in 1948, but its relatively high cost prevented any widespread use.

model. It was a glass-covered box containing several cylindrical water tanks made of blackened aluminum or copper. The heater held an average of 85 gallons and was usually installed on a slanted platform or on the south-sloping roof of the house. Connected to the city water lines, the heater reached its maximum temperature by mid-afternoon. Because of its increased heat collection in winter, it produced more hot water than earlier models during the course of a year.

Before long, however, the aluminum tanks corroded and developed leaks. Manufacturers turned to other cheap, durable materials such as glass, stainless steel and plastic. Because city dwellers generally used more hot water than people in the countryside, they often coupled two or three water heaters together. At about $50 each, this tank-type heater was a far more expensive proposition than the soft vinyl or wood-basin heaters. Nevertheless, they still sold extremely well; in 1960 alone over 15,000 of these heaters were sold. Sales increased to 50,000 the following year, and to over 200,000 by 1963. Over the next five years, the combined sales by the five major manufacturers of tank-type heaters averaged over a quarter of a million units per year. The soft vinyl heater continued to hold its own in rural districts, itself averaging over a quarter of a million sales annually during the same period. As a result, total sales of solar water heaters in Japan reached 3.7 million by 1969. The fuel savings resulting from such widespread use was estimated to be nearly 50 million tons of fuel per year.

But 1966 turned out to be the peak year for sales. During the 1960's Japan began to import huge quantities of oil from the Middle East and Indonesia. Cheap, available fossil fuels drew many customers away from solar heaters. Rural electrification gave farmers another alternative to rice straw for water heating, as many took advantage of low nighttime electric rates. In addition, an increasing number of middle-class families preferred a domestic water-heating system tied into the central house-heating system—a new feature in urban Japanese homes and apartments.

The tank-type solar water heaters experienced many of the same problems as had the early Climax heaters. None of them had insulated storage tanks, so all lost large quantities of heat at night. Furthermore, the relatively large volumes of water in the tanks didn't get really hot until mid-afternoon or later. To obviate these shortcomings, Professor Tanashita had introduced a system with a collector and separate, insulated storage tank as early as 1948. But few could afford its high price in postwar Japan, and the simpler versions held sway. Finally in the late 1960's, cheap energy and a desire for more convenience brought a sharp drop in sales of all solar water heaters. As elsewhere, the industry eventually died out.

Solaron collectors heat this duplex home in Fort Collins, Colorado.

Chapter 19
Full Circle

Time after time, scarcities of fuel have stimulated the search for energy alternatives—spurring advances in solar architecture and technology. But when people discovered abundant new sources of fuel, solar energy became "uneconomical" and dropped from sight. In California and Florida, solar water heaters remained immensely popular until cheap gas or electricity displaced them. Sometimes no alternative fuels were found. Such was the case in ancient Greece and Rome, where wood shortages only worsened, and people continued to use solar architecture for centuries.

Today's energy shortages resemble those of ancient Greece and Rome. Conventional fuels are becoming scarce and expensive, and alternatives are limited. The extraction of oil from tar sands or shale and the production of synthetic fuel oil from coal are complex processes that involve a host of economic and ecological problems. Nuclear power, once touted by many scientists and engineers as the ultimate solution to our energy problems, is undergoing an intensely critical reevaluation. Escalating construction costs, the lack of a safe waste disposal system, the potential for catastrophic disaster as demonstrated in the accident at Three Mile Island, and the danger that atomic fuel will be converted into weapons make the nuclear dream look more like a nightmare. Breeder reactors, the second generation of nuclear fission power plants, present far more serious problems. Fusion promises unlimited power, but it may never attain technical feasibility.

The sun, however, is a proven energy source that can meet many of our energy needs indefinitely. Solar architecture and technology have been evolving for more than 2,000 years. Many of the present applications resemble those tried earlier, but future improvements in materials and manufacturing processes promise further innovations and even more widespread use.

There is an almost unanimous consensus among experts about the importance of solar orientation and building design. Early in the 1970's a new generation of solar single-family dwellings was built—mostly in the American Southwest. A growing number of architects and urban planners are just beginning to realize that designing entire solar communities makes even more sense. As in ancient Greece, in Germany during the Weimar Republic, and in the United States during the 1940's, large solar housing developments based on rational land use have begun to reappear. The Village Homes subdivision built at Davis, California, demonstrates the practicality of solar housing in a community designed for energy conservation. Solar architecture appeals to many because it requires no major breakthroughs and only a small additional capital outlay. If a society implemented this strategy on a mass scale, it would realize enormous savings in fossil fuels.

After the oil embargo of 1973-1974, many people around the world started using simple and practical flat-plate collectors for hot water and home heat. Israel now leads the world in per capita use of the sun; a third of all Israeli homes get their hot water from solar collectors. Australians have bought over 200,000 flat-plate collectors, and the Japanese solar heater business is booming, too. In the United States there have been a tremendous number of new solar installations during the last several years. In the Friends' Community Development near Boston, Massachusetts, all of the townhouses obtain about 50 percent of their winter heat from individual solar air-heating collectors.

Once again the sun is powering water pumps for irrigation and other tasks in

Above: Aerial view of the Village Homes subdivision in Davis, California. Almost all the homes receive a majority of their heat from the sun.

Above left: The Kita-Kyushu swimming complex near Tokyo, Japan. Beasley collectors made by the Azuma-Koki Company keep the pool water warm.

Left: The Neptune Hotel in Israel's resort city of Eilat uses flat-plate collectors to heat its water supplies.

remote, sunny areas. As at the turn of the century, there is a controversy over the best approach—whether high or low-temperature collection devices are best suited to drive these pumps. In Willard, New Mexico, a parabolic-trough collector similar to Ericsson's was installed in 1977 to drive irrigation pumps. The sun powers a 25-horsepower engine that pumps almost 700 gallons of water per minute and irrigates 404 acres of crops. Others are using solar reflectors for generation of electricity. Debate rages over whether to use such devices on a large or small scale.

Silicon solar cells of the type developed by Bell Telephone Laboratories in the 1950's are being used more frequently as ways are found to manufacture them cheaply. Their cost has dropped from more than $2,000 per peak watt in 1960 to about $10 per watt at this writing. Other kinds of solar cells look promising, too, especially those made from amorphous, noncrystalline silicon. Many scientists predict that individual homes and buildings will someday have their own solar-cell power supplies.

The sun can serve as a practical, bountiful energy source that will sustain civilization when current fossil fuel supplies run out. There is little reason to consider solar energy as an "exotic, unproved" technology whose practical application is years away. The remaining economic barriers to its widespread use are

Parabolic-trough collectors furnished by the Acurex Corporation are used in this solar pumping experiment in Willard, New Mexico.

rapidly eroding as conventional fuels become scarce and expensive. Perhaps we are now at the threshold of an enduring solar age.

History offers many lessons that can smooth our transition to this new age. The successes and failures of past generations can help guide us in the development of solar applications. But the most important lesson is that solar energy can be a practical alternative to scarce fossil fuels. The Roman wood ships navigating the Mediterranean have been replaced by oil tankers bound for the Persian Gulf. Their quest remains the same. But the sun still beats down upon us even though the forests of North Africa have long since disappeared. And it will beat down on future generations long after all the oil and gas wells are depleted.

A Golden Thread is based on extensive research that used varied and sometimes obscure sources. The following notes are intended to direct inquiring readers to these sources for closer examination. Occasionally, the notes elaborate upon ideas presented in the text.

The numerals listed to the left of these notes indicate the page or pages of the text where the corresponding source material is used or quoted. Only the first occurrence of a particular page number is given; subsequent unnumbered notes refer to the same page. When a source is used on one (or more) following page(s) in the text, the symbol "f." (or "ff.") appears after the page number.

Chapter 1: Solar Architecture in Ancient Greece

3 For Socrates' comments on the ideal house, see Xenophon, *Memorabilia* III, viii, 8 f.

Concerning the protection of olive groves, see K. Sklawounos, "Uber die Holzversorgung Griechenlands im Altertum," in *Forstwissenschaftliches Centralblatt*, Vol. 52 (1930), p. 273.

Each year the Athenians could observe the continual recession of their forests. By the fifth century B.C., oak—the preferred source of charcoal—had been depleted as far as Mount Parnes, more than ten miles from Athens. See Aristophanes, *Acharnians*, 666. The destruction of Athenian forests must have proceeded at an alarming rate for Plato to have written that these environs had become so bare that they could only " support bees" where once lush woods had stood. Due to such deforestation, Plato told his readers, erosion was now rampant, causing the land to lose "the water which flows off the bare earth into the sea." In former times, he continued, the dense foliage allowed "the land to reap the benefit of the rain." See Plato, *Critias*, III—B-D.

Concerning the Delian regulations on charcoal sales, see: E. Schulhof and P. Huvelin, "Loi Reglant la Vente du Bois et du Charbon à Delos," in *Bulletin de Correspondence Hellenique*, Vol. 31 (1907), pp. 47-93; Auguste Jarde, "Note sur une Inscription de Delos," in the same *Bulletin*, Vol. 47 (1923), pp. 301-306; and Phillipe Gauthier, "Les Ventes Publiques de Bois et Charbon: A Propos d'une Inscription de Delos," in the same *Bulletin*, Vol. 101 (1977), pp. 203-208.

Concerning the ban on exportation of locally produced Athenian wood, see August Böckh, *Staatshaushaltung der Athener* (Berlin: Druck and Verlag von Georg Reimer, 1886), Vol. 1, p. 68.

Delos, for example, totally lacked any indigenous fuel supplies. See Schulhof and Huvelin, p. 60.

Theophrastus quoted from Theophrastus, *De Igne*, 6.

3 f. Oribasius' quote on health and sunlight from Oribasius, II, 5.317.

4 According to Herodotus ii.109.3, the Babylonians introduced the *gnomon*. On telling time with this device, see Athenaeus, *Deipnosaphistae*, 1.7-8.

Notes

Aristotle quoted from his *Economics*, I, 6.7.

Socrates quoted in Xenophon, *Memorabilia*, III, viii, 8 f.

5 ff.　For the most authoritative analysis of Olynthian architecture, see David M. Robinson and J. Walter Graham, *Excavations at Olynthus*, 14 volumes (Baltimore: Johns Hopkins Press, 1929-1952). See especially Vol. 8, "Domestic Architecture."

6　Aristotle's quote about "the modern fashion" from his *Politics*, VII, 11.6. Aristotle credits the famous urban planner Hippodamus with introducing the "checkerboard" street plan.

The use of sun-dried adobe bricks at Olynthus suggests another Greek application of solar energy for fuel conservation. According to Eugene Ayres and Charles Scarlott, *Energy Sources–The Wealth of the World* (New York: McGraw-Hill, 1953), p. 9, each cubic foot of sun-dried brick used instead of burnt brick saved 150 cubic feet of wood.

10　Controversy surrounds this quote attributed to Socrates in Xenophon's *Memorabilia*. In the most commonly available translation, the Loeb series, Socrates is credited with saying, "We should build the south side loftier to get the winter sun and the north side lower to keep out the north winds." Robinson and Graham contend that this passage has been mistranslated and should read, "The southern part of the house is to be built lower than the northern in order not to cut off the sun; whereas the northern part is to be built higher than the southern in order to exclude the cold north winds." See Robinson and Graham, pp. 145-146. We have accepted their interpretation.

10 f.　For more information on the original excavations at Priene, refer to T. Weigand and H. Schrader, *Priene* (Berlin: Georg Reimer, 1904). See especially chapter 10, "Die Privathauser."

11　The original excavations of Delos appeared in École Française d'Athenes Exploration Archeologique de Delos, 21 volumes (Paris: Éditions de Boccard, 1909-1959). See especially Joseph Chamonard, "Le Quartier du Théatre: Étude sur l'Habitation Delienne à l'Epoque Hellenistique," Vol. 8, Part 1, 1922-24.

As to the opinions of the excavators of Olynthus, Priene and Delos about the importance of building orientation in capturing solar heat, see: Robinson and Graham, pp. 144-145; Weigand and Schrader, p. 290; and Chamonard, p. 141.

13　For studies of ancient Chinese architecture see: Andrew C. Boyd, *Chinese Architecture and Town Planning* (Chicago: University of Chicago Press, 1962); Dennis G. Mirams, *A Brief History of Chinese Architecture* (Shanghai: Kelly and Walsh Ltd., 1940); and Nelson Wu, *Chinese and Indian Architecture* (New York: Brazillier, 1963).

Isomachus is quoted in Xenophon, *Economics,* 9.4.

Thatcher's solar heating data comes from his study: Edwin D. Thatcher, "The Open Rooms of the Terme del Foro at Ostia," in *Memoirs of the American Academy in Rome,* Vol. 24 (1956), pp. 167–264.

Aeschylus wrote this description of the barbarians in his *Prometheus Bound,* 447 f.

Chapter 2: Roman Solar Architecture

15   Kretschmer reported these results on Roman *hypocausts* from a test he conducted using five furnaces. For details of this study, see R.J. Forbes, *Studies in Ancient Technology* (Leiden: E.J. Brill), Vol. 6 (1966), pp. 55-56. Concerning Kretschmer's original work, see F. Kretschmer, "Der Betriebsversuch an einem Hypokaustum der Sallburg," in *Germania*, Vol. 31 (1953), pp. 64-67, and *Saalburg Jahrbuch*, Vol. 12 (1953), pp. 8-41.

Producing a ton of copper per day, for example, would require an acre of trees. For details, see Forbes, p. 19. Combining the demands made on forests to supply fuel for bath heating, house heating and industry, and wood for home and ship construction, one can begin to understand why wood became scarce in many areas of the Roman Empire.

The description of the forests in Mount Cimino can be found in Livy, *Histories*, 9.36-38.

Vergil wrote in his *Geogica*, 2.440-445, that Caucasian timber was used for home and ship building.

For details of the fuel crisis on Elba, see Strabo, *Geography*, 5.2.6. Pliny also wrote of fuel shortages in the Campania region. See his *Natural History*, XXXIV.xx.96. Concerning fuel shortages in the entire Roman world, see George P. Marsh, *The Earth as Modified by Human Action* (New York: Scribner, Armstrong and Company, 1847), p. 320; also see Ernst Pulgram, *Tongues of Italy* (Cambridge: Harvard University Press), pp. 35-37. Marsh relates the changing Roman methods of brick preparation to the increasing scarcity and cost of fuel.

Vitruvius is quoted from his treatise, *On Architecture*, VI.i.1.

Concerning building in temperate climates, see Vitruvius, VI.i.2. For the orientation of different rooms, refer to Vitruvius, VI.iv.1-2.

17   Concerning the homes at Herculaneum and their orientation, see Amadeo Mauri, *Ercolano* (Rome: Instituto Poligraphico dello Stato, 1927-28), especially Vol. 1, pp. 380-302.

17 ff.   Pliny wrote about his two villas in his *Letters*, II.17 (Laurentum) and V.6 (Tuscany). John Howell Westcott, in his *Selected Letters of Pliny* (Norman: University of Oklahoma Press, 1965), comments on p. 167 that "much dependence was placed on the sun for heating interiors in cold weather." In Adrian M. Sherwin-White's *The Letters of Pliny* (Oxford: Clarendon Press, 1966), the author states, "The Romans attached great importance to the correct seasonal orientation of rooms and houses since warmth depended, as these letters indicate, much more upon aspect than upon . . . the heating system," p. 191.

19   The word *heliocaminus* is derived from two Latin words—*helio* (solar) and *caminus* (furnace). Regarding the definition of *heliocaminus*, see Aegidio Forcellini, *Lexicon Totius Latintatis*, Vol. 2 (1940), p. 645. Charlton Lewis and Charles Short, in their authoritative work, *A Latin Dictionary* (London: Oxford University Press, 1879), p. 845, define a *heliocaminus* as "an apartment exposed to the sun, used as a winter abode."

An excellent source on the topic of Roman glassmaking is Forbes, Vol. 5, pp. 185-187. On the methods of making window glass see D.B. Harden, "Domestic Window Glass: Roman, Saxon and Medieval," in E.M. Jope, *Studies in Building History* (London: Odhams Press, 1961), pp. 41-42. For shaping window panes out of transparent stone, refer to Pliny, *Natural History*, XXXVI.x1v.160.

Although Seneca in *Epistulae Morales*, XC.25 did not specifically say that these window panes were made from transparent stone or glass, it can be safely conjectured that he was talking about glass. He referred to the substance from which the panes were made as *"testa,"* Latin for "a baked material." The only known " baked material" that is transparent and used for window panes was glass. It was not until 300 years later that the Romans differentiated windows made from stone from those made with glass. They used the general term *"specularium"* to refer to either material. See Hugo Blumner, *Technologie und Terminologie der Gewerbe und Kunste bei Griechen und Romern* (Leipzig und Berlin: B.G. Teubner, 1912), p. 66. Pliny, in his *Natural History*, XXX.xlvi.163, stated that the "specularium" allowed sunlight to enter the building. Seneca also said it "admits clear light."

On Tiberius' penchant for cucumbers and his placing them under transparent coverings, see Pliny, *Natural History*, XXIII.xxiii.64.

20 Concerning glassed cold frames, Martial in his *Epigrams*, 8.68, tells us that a friend Entelles shut his vineyard in transparent glass so "that the jealous winter may not sear the purple clusters nor chill frost consume the gifts of Bacchus."

Martial's satiric complaint is found in his *Epigrams*, 8.14.

21 Seneca's account of activity in the bath is found in his *Epistulae Morales*, LVI.1-2.

Seneca's comments on Roman baths in general are from his *Epistulae Morales, LXXXVI.6-11.*

On the orientation of Roman baths, see Vitruvius, V.x.i and VI.iv.1.

On the use of solar heat in Roman sweat rooms, see Atti dell R. Accademia Nazionale dei Lincei, *Notizie Degli Scavi di Antichita*, Vol. 19 (1922), pp. 241-245.

25 For a comparison of the old and new baths at Pompeii, see August Mau, *Pompeii* (London: Macmillan, 1902), pp. 208-211.

Concerning wood importation from North Africa to Rome see the *Codex Theodesius* 13.5.1.10. In this passage, the Emperors of Rome order the African proconsul to grant special privileges to African shipmasters who would carry wood from Africa to Rome. Quintus Aurelius Symmachus talked of similar special dispensations to shipmasters in his *Q. Aurelii Symmachi Relationes*, Recensut Guilrelmers Meyer (Lipsiae: B.G. Teubneri, 1872), Relato 44. Much of this wood was used for heating purposes; see J.P. Waltzing, *Étude Historique sur les Corporations Professionales Chez les Romains* (Bologna: Forni Editore, 1968) Vol. II, p. 55 and pp. 125-126.

25 f. Concerning the economic conditions that prevailed during the later Roman Empire, see F. Oertel, "The Economic Life of the Empire," in S.A. Cook et al., *The Cambridge Ancient History* (Cambridge: University Press, 1939), Vol. 12, pp. 232-281.

26 f. For a discussion of the role that Faventinus and Palladius fulfilled for their rural aristocratic clients, see William H. Plommer, *Vitruvius and Later Roman Building Manuals* (London: Cambridge Press, 1973), pp 2-3.

For the recommendations of Faventinus and Palladius, refer to: M. Ceti Faventini, *De Diversis Architectionicae*, 16, and Rutillus T. Palladius, *De Re Rustica*, 1.40 on the positioning of baths; Faventini, 14 on the location of winter rooms; Faventini, 26, Palladius, 1.40, and also Vitruvius, VII.iv.4 on the use of rubble for absorbing solar heat; Palladius, 1.42 and 1.40 on the recycling of bath water and the use of waste heat; and Palladius, 1.20 on solar heat traps to keep oil from congealing in winter.

27 On Roman sun rights laws, see Ulpian, *Digestia* 8.2.17.

Chapter 3: Burning Mirrors

29 Concerning the evolution of mirror technology in ancient times, Al Singari, a medieval Moslem mathematician, wrote: "the ancients made mirrors of plane surfaces. Some made them spherical until Diocles showed and proved that if the surface of these mirrors had its curvature in the form of a parabola, they then have the greatest power and burn most strongly." See Sir Thomas Heath, *A History of Greek Mathematics* (Oxford: Clarendon Press, 1921), p. 201. Diocles gave Dositheus the credit for building the first parabolic mirror in his *On Burning Mirrors*, translation and commentaries by C.J. Toomer (Berlin; New York: Springer, 1976), p. 34.

Probably no myth has persisted as fact for so long as the story of Archimedes burning the Roman fleet with the aid of mirrors. This tale began with Galen, writing several hundred years after Archimedes' alleged feat: "I believe they say that Archimedes set fire to the enemy's ships by means of burning mirrors." See Galenus, *Opera Omina* (Hildesheim: G. Olms, 1964-1965), Vol. 1, p. 657. From this modest speculative sentence, later writers speculated about how Archimedes' feat was performed. For a thorough repudiation of the myth, see Hugo Blumner, *Technolgie Und Terminologie Der Gewerbe Und Kunste Bei Griechen Und Romern* (Leipzig und Berlin: B.G. Teubner, 1912), p. 36.

Plutarch discussed the Vestal Virgins' use of mirrors in their religious rites in his *The Lives of the Noble Grecians and Romans*, translated by John Dryden and revised by Arthur Hugh Clough (New York: The Modern Library), p. 82.

On the Chinese use of burning mirrors see Joseph Needham, *Science and Civilization in China* (Cambridge: University Press, 1954), pp. 87-89.

30 Al-Haitham quoted in Ibn Al Haitham, *A Discourse on the Parabolodial Focussing Mirror*, translated by H.J.J. Winter and W. Arafat, in *The Journal of the Royal Asiatic Society of Bengal*, Series 3, Vol. 15, Nos. 1 & 2 (1949), p. 25.

30 f. For Bacon's quotes concerning the Anti-Christ see *The Opus Majus of Roger Bacon*, translated by Robert B. Burke (Philadelphia: University of Pennsylvania Press, 1928), p. 135. On the building of mirrors and their availability as weapons in defense of Christendom see Roger Bacon, *Fr. Rogeri Bacon Opera*, edited by J.S. Brewer (London: Longman, Green, Longman, and Roberts, 1959), *Opus Tertium, XXXVI*.

32 f. Concerning Leonardo's plans for the construction and use of his giant mirror see *Il Codice Atlantico di Leonardo da Vinci nella Bibliotecta Ambrosiana Milano* (Milani: U. Hoepli, 1894-1904), 371 v-a (blue paper) and 277 r-a. For further discussion about da Vinci's giant mirror see Carlo Pedretti, *The Literary Works of Leonardo da Vinci* (Berkeley: University of California Press, 1973), Vol. 2, pp. 19-20. On Verocchio's use of mirrors for soldering metals, see Pedretti, p. 120.

33 Lonicier is quoted in Augustin Mouchot, *La Chaleur Solaire* 2nd ed. (Paris: Gauthier-Villars, 1879), pp. 94-95.

34 For a description and discussion of Magini's mirror see Giovanni Magini, *Breve Instruttione Sopra L'Apparenze et Mirabili Effetti dello Specchio Concavo Sferico* (Bologna: Presso Clementi Ferroni, 1628), Ch. 3.

34 ff. Concerning sixteenth, seventeenth and eighteenth-century mirror construction techniques see Mouchot, p. 108.

35 Galileo expressed this opinion through the voice of a fictional character Sagacious. See Galileo Galilei, *Dialogue Concerning the Two New Sciences*, translated by Henry Crew and Alfonso de Salvio (Evanston and Chicago: Northwestern University Press, 1950), p. 43.

35 f.   For Porta's criticism of Cardano, his claim of building the ultimate mirror, and his discussion
        of the destructive potential of focussing devices, see G. Porta, *Natural Magick* (London: for
        Thomas Young and Samuel Speed, 1658), Ch. 17, Sections 15, 17, and 9. Encyclopedias and
        catalogs list Porta's first name as either Giambattista or Giovanni; we have used the former.

37   Athanasius—Kircher reported his investigations in his *Ars Magna Lucis Et Umbrae*
        (Romae: Sumptibus Hermanni Scheus, 1646), Part 3, Book 10, Corollary 2.

37 f.   Accounts of Villette's experiments were recorded in *The Philosophical Transactions of The
        Royal Society of London*, Vol. 1, No. 6 (1665); Vol. 4, No. 47 (1669); and Vol. 30, No. 360
        (1719).

38   The report and criticism of Gartner's work appeared in Peter Hoesen, *Kurze Nachricht Derer
        Beschaffenheit Und Wirkung Derer Parbolischen Brennspiegel* (Dresden; Bey F. Hetel,
        1755), Chapter 1.

        Baron Tchirnhausen's report was published in *The Philosophical Transactions of The Royal
        Society of London*, Volume 16, No. 188 (6887).

38 f.   For a detailed discussion of Hoesen's work see Hoesen, pp. 1-15.

Chapter 4: Heat for Horticulture

40 ff. Probably the best historical account of the evolution of the greenhouse can be found in John
        Hix, *The Glass House* (Cambridge: M.I.T. Press, 1974).

41   In England, for example, glass windows continued to be a luxury until the latter part of the
        sixteenth-century. See Eleanor Godfrey, *The Development of English Glassmaking, 1560-
        1640* (Chapel Hill: University of North Carolina Press, 1975), p. 207. Likewise, residents in
        many important cities of sixteenth-century Europe were still using linen, paper, and parch-
        ment panes. See R.J. Forbes, *Studies in Ancient Technology* (Leiden: E.J. Brill, 1966),
        Volume 5, p. 185.

        For Loudon's quote on horticulture in the low countries see John C. Loudon, *Remarks on the
        Construction of Hot Houses* (London: for J. Taylor, 1817), p. 3.

42   From 1550 to 1850 the greatest advances of Northern Hemisphere glaciers occurred since the
        Ice Age. See H.H. Lamb, *The English Climate* (London: English Universities Press, 1964),
        pp. 162-164. Another good source on English weather during these centuries is J.H. Brazell,
        London Weather (London, 1968), pp. 8-10.

        The quote about the gardener's fight against the cold can be found in John Laurence, *The
        Clergyman's Recreation* (London: for B. Lintott, 1726), p. 72.

        Carpenter quoted in Joseph Carpenter, *The Retir'd Gardener* (London: for J. Tonson, 1717).

43   On the use of brick for its heat absorbing qualities, refer to Nicolas Fatio de Dullier, *Fruit
        Walls Improved* (London: R. Everingham, 1699), p. 3. Also see Stephen Switzer, *Ichnog-
        raphia Rustica* (London: D. Browne, 1718), p. 239.

        Fatio de Dullier's quote about the shortcomings of vertical fruit walls is in Fatio de Dullier,
        p. 2.

        Concerning Switzer's advocacy of perpendicular fruit walls facing southeast see Stephen
        Switzer, *The Practical Gardener* (London: for T. Woodward, 1724), p. 295. "Half-rounds"
        are discussed in Switzer's *Ichnographia Rustica*, p. 240.

43 ff. Fatio de Dullier's calculations on sloping fruit walls and his predications of better crops are to be found in Fatio de Dullier, pp. 4-6.

47   The Duke of Rutland's sloping walls had to be heated artificially, too. See Switzer, *The Practical Fruit Gardener*, p. 305.

Platt quoted in Sir Hugh Platt, *The Garden of Eden* (London: for William Leake, 1659-1660), p. 41.

For an excellent discussion of the evolution of flat glass technology refer to Frances Rogers and Alice Beard, *5000 Years of Glass* (New York: Fredrick A. Stokes, 1937), pp. 140-150. See also R. W. Douglas and Susan Frank, *A History of Glassmaking* (Henley-On-Thames, Oxfordshire: Foulis, 1972), pp. 137-148.

Concerning fuel savings as well as establishing a more healthful environment for plants by relying on solar heat see James Anderson, *A Description of a Patent Hot-House* (London: for J. Cumming, 1803), pp. vi-vii.

About Boerhaave's calculations see Hermann Boerhaave, *Element Chemiae* (Londini: Sumtibus S.K. et J.K., 1732) Vol. 1, p. 213.

48   On greenhouses designed for optimum summer orientation in England see Knight, "Description of a Forcing House for Grapes," in *London Horticultural Transactions,* Vol. 1, p. 99.

48 f. Adanson's greenhouse plans are to be found in Michael Adanson, *Familles des Plantes,* (Paris: Vincent, 1763), Vol. 1, p. 132.

49   Insulating the greenhouse at night or during inclement weather is discussed in Phillip Miller, *The Gardener's Dictionary* (London, 1937).

49 f. The description of Anderson's novel method of solar heat storage is taken from Anderson, pp. 23-42.

50   The difference between a greenhouse and a conservatory is pointed out by Robert Kerr, *The Gentleman's House* (London: J. Murray, 1864), p. 353.

Concerning fuel savings and the proper orientation of conservatories refer to John C. Loudon, *Encyclopedia of Cottage, Farm, and Villa Architecture*, (London: F. Warne & Company, 1846), p. 974.

51 f. *The English Gardener* quoted in Hix, p. 89.

52   Hix quote is taken from Hix, p. 87.

The versatility of the attached conservatory is discussed by Kerr, p. 141.

Replacing dark parlors with rooftop conservatories was the idea of W. Bridges Adams, who was quoted in *The Gardener's Chronicle* (January 14, 1860), p. 29.

Jacob Forst quoted in *The Gardener's Chronicle* (January 14, 1860), p. 29.

Chapter 5: Solar Hot Boxes

55   The original account of de Saussure's experiments with hot boxes appeared in his letter to the editors of *Le Journal de Paris*, Supplement Number 108 (April 17, 1784), pp. 475-478. De Saussure expressed his uncertainty about why his hot box heated up in his *Voyages Dans Les*

Alpes (Geneva: Barde, Manget, and Compagnie, 1786), 1662, Vol. 2, Ch. 35, paragraph 933.

56 f. How the hot box served as a model of our atmosphere was discussed in Svante Arrhenius, *Worlds in the Making* (New York: Harper, 1908), p. 51.

57 f. Herschel's account appeared in Sir John F.W. Herschel, *Results of Astronomical Observations Made During the Years 1834, 5, 6, 7, 8, at the Cape of Good Hope* (London: Elder and Company, 1847), pp. 443-444.

   48 Herschel wrote of his solar cook-out with his sons in his diary dated December 4, 1837, as quoted in David Evans, et al., *Herschel at the Cape* (Cape Town: Balhemma, 1969), p. 330.

      Langley wrote of his childhood fascination with glass solar heat traps in a letter dated June 23, 1887. He stated that "One of my very earliest childish experiences is connected with my enquiries as to the reason that glass in a hot bed I saw kept the contents warm."

   59 For a detailed account of Langley's experiments with the hot box on Mount Whitney, see Samuel P. Langley, *Researches on Solar Heat* (Washington: U.S. Government Printing Office, 1884), pp. 166-168.

Chapter 6: The First Solar Motors

   63 Augustin Mouchot expressed his concern about his country's fuel situation in his book, *La Chaleur Solaire,* 2nd ed. (Gauthier-Villars: 1879), p. 256. Mouchot pointed out that wind, water, and garbage could also be used to generate power. This book chronicles the history of solar devices up to that time and documents all of Mouchot's work. Unless otherwise cited, all quotes from Mouchot come from *La Chaleur Solaire*.

      The quote on early solar devices is from Mouchot, p. 154.

      Concerning Hero's solar syphon see Hero of Alexandria, *The Pneumatics of Hero of Alexandria*, Introduced by Marie B. Hall (London: Macdonald, 1971), p. 69.

      Kircher's solar clock is discussed in Mouchot, pp. 170-178.

   64 The solar whistle was built by Solomon de Caus and appears in Issac de Caus, *New and Rare Inventions of Water Works* (London: Joseph Moxon, 1659), p. 33.

      The quote on practical applications of solar is from Mouchot, p. 261.

   65 The bell-shaped solar collector is discussed in Mouchot, p. 267. His quotes appear on p. 123. Also see Augustin Mouchot, "Sur les Effects Mecaniques de l'Air Confine et Chauffe par les Rayons du Soleil," *Comptes Rendus de l'Academie des Sciences,* Vol. 59 (1864), p. 527.

   66 For details of the solar oven, see Mouchot, pp. 142-143.

67 f. Details of Mouchot's first solar motor appear in Chapter 6 of *La Chaleur Solaire*. His words describing the initial test run are found on p. 206.

   69 Concerning the disappearance of the solar motor during the Franco-Prussian conflict see Mouchot, p. 224.

      The description of the frustrated attempts at getting funding can be found in Mouchot, p. 225.

69 f. Leon Simonin discussed the Tours motor in "L'Emploi Industriel de la Chaleur Solaire," *Revue des Deux Mondes*, May 1, 1876.

70 Mouchot credits Dupuy as the source of his conical mirror shape in *La Chaleur Solaire*, Note A.

Bystanders' quotes come from Simonin. For a further description of the Tours motors see Augustin Mouchot, "Resultats Obtenus dans les Essais d'Applications Industrielles de la Chaleur Solaire," *Comptes Rendus de l'Academie des Sciences*, Vol. 81 (1875), pp. 571-574. Also see Mouchot, pp. 222-230.

The comments about using sun motors in cities is from Mouchot, p. 257.

The quote concerning sun power in the French colonies appeared in Mouchot, p. 263.

71 The quote concerning portable solar ovens for the African troops appeared in Mouchot, pp. 263-264. Specific examples of the other experiments in Algeria are found in Mouchot, "Re'suitats d'Experiences Faites en Divers Points de l'Algerie pour l'Emploi Industriel de la Chaleur Solaire," *Comptes Rendus de l'Academie des Sciences*, Vol. 86 (1878), pp. 1019-1021.

72 The comments on distilling and pasteurizing wine are from Mouchot, pp. 235-236. Because wine and liquors were important exports, the French government was very interested in the practicality of solar distillation.

The Paris Exposition motor is described in Mouchot, pp. 243-246.

Ferdinand Carré's refrigeration system used a solution of ammonia and water as the heat transfer fluid. The ammonia separated from the water when heated, and circulated through a condensor. The liquid ammonia then was allowed to evaporate, causing refrigeration. Mouchot's quote on this refrigeration experiment is found on p. 263 of *La Chaleur Solaire*. He was apparently the first person to use solar energy to provide the heat needed in Carré's apparatus—the first solar absorption cooling device.

73 The discussion of solar thermoelectrical generation is found in Mouchot, pp. 248-249. Mouchot exhibited exceptional foresight in suggesting that the products of electrolysis could serve as a storage system for solar electricity as this method is currently touted as one of the most promising. Mouchot employed a Clamond thermoelectric battery which operated on principles discovered in 1822 by Thomas Seebeck.

74 The French government study of Mouchot and Pifre's sun motors is from A. Crova, "Etude des Appareils Solaires," *Comptes Rendus de L'Academie des Sciences*, Vol. 94 (1882), pp. 943-945. This paper gives only a few figures from the text.

On Pifre's sun-powered printing press and other work, consult A. Pifre, "Nouveaux Résultats d'Utilization de la Chaleur Solaire Obténus à Paris," *Comptes Rendus de l'Academie des Sciences*, Vol. 91 (1880) pp. 388-389. See also "A Solar Printing Press," *Nature* (Sept. 21, 1882), pp. 503-504.

Chapter 7: Two American Pioneers

77 For Ericsson's comments on producing salt by solar evaporation, see William C. Church, *The Life of John Ericsson*, Vol. II (New York: Scribners, 1890), p. 266. Church's book is the most complete account of Ericsson's life and accomplishments.

Ericsson's comments on the potential of solar energy are in Church, p. 266. His original study on solar energy was presented in 1868 to the University of Lund (Sweden), entitled "The Use of Solar Heat as a Mechanical Motor-Power." He received an honorary doctorate for this paper.

The remark about Mouchot's sun motor is from John Ericsson, *Contributions to the Centennial* (New York: 1876), p. 564. Ericsson held firmly to the belief that he had invented the first solar motor.

The pledge not to divulge specifics about his inventions comes from Ericsson, p. 559.

78   A detailed description of the solar hot-air engine appears in Ericsson, pp. 571-574. The motor closely resembled the popular engine developed by the Scotch clergyman the Reverend Robert Stirling.

Ericsson's letter to Harry Delameter was written on October 22, 1873 and is quoted from Church, p. 268.

78 f.   Ericsson's discussion of the cost of solar-powered motors appears in Church, p. 271. His recommendation for siting sun motors in sunny regions can be found in Ericsson, p. 563.

79 f.   Information about the inexpensive reflector and sun motor can be found in John Ericsson, "The Sun Motor," *Nature*, Vol. 29 (January 3, 1884), pp. 217-218 and Vol. 38 (August 2, 1888), pp. 319-321.

80   The announcement of the perfected sun motor appeared in Ericsson, "The Sun Motor," *Nature*, Vol. 38 (August 2, 1888), p. 321. It is not clear why the commercial production of his machine did not commence between the time of this announcement and his death.

Ericsson's obituary from *Science* is quoted in *Nature*, Vol. 39 (March 28, 1889), p. 517.

81   The *Harper's* quote is from "Power from Sunlight," *Harpers Weekly*, Vol. 47 (February 4, 1903), p. 256.

Pope's quote is from Charles Pope, *Solar Heat* (Boston: 1903), p. 13. Pope's book is the first text written in English to cover the history and uses of solar energy.

Thurston's quotes are from Robert H. Thurston, "Utilizing The Sun's Energy," *Smithsonian Institution Annual Report* (1901), p. 265.

Pope's criticism of many of the patents taken out on solar devices appears on p. 33 of *Solar Heat*.

82   Eneas' quotes concerning his efforts to find literature on solar motors and details of initial work are from correspondence between the inventor and Samuel Pierpoint Langley, Secretary of the Smithsonian Institution, *The Smithsonian Archives*, Record 31 (Office of the Secretary: 1891-1906), Incoming Correspondence, Box 23.

82 f.   Technical details of Eneas' first invented conical reflector and sun motor are from U.S. Patent 670,916 (March 26, 1901).

84   Information on Eneas' improved conical reflector support is from U.S. Patent 670, 917 (March 26, 1901).

85   Anna Laura Meyers, who lived in Pasadena at the turn of the century, spoke of the ostrich farm and the solar motor in a private interview.

Magazine quotes on the Pasadena motor are from the following sources: Arthur Inkersley, "Sunshine as Power," *Sunset*, Vol. 10 (April, 1903), p. 515; Frank B. Millard, "Harnessing the Sun," *World's Work*, Vol. 1 (April, 1901), p. 600; Alfred L. Davenport, "The Solar Motor at Pasadena, California," *Engineering News*, Vol. 45 (May 9, 1901), p. 330.

Quote regarding the extreme heat along the focus of Eneas' mirror is from C. Holder, "Solar Motors," *Scientific American*, Vol. 84 (March 16, 1901), p. 169.

88 The Estrada quote is from a private interview.

Quotes and background information on Arizona sun motor were obtained from Ruth Mellenbruch, "Solar Energy Experiment Was Conducted Here in 1906-1907," *Arizona Range News* (September 12, 1941).

89 The cost of Eneas' motor was obtained from "The Eneas Solar Water Pump," *Sun at Work*, Vol. 9 (Fourth Quarter, 1965), pp. 12-14.

Eneas' criticism of reflectors to power sun machines comes from correspondence between Eneas and Langley and can be found in *The Smithsonian Archives*, February 10, 1906.

Chapter 8: Low Temperature Solar Motors

91 Background information on Tellier is from *The Encyclopedia La Rousse* (Paris: Librarie Larousse, 1932), Vol. 6, p. 627.

94 Our primary source of information on Tellier's first solar pump is from "The Utilization of Solar Heat for The Elevation of Water," *Scientific American*, Vol. 53 (October 3, 1885), p. 214.

Technical details on Tellier's first pump comes from *Kaiserliches Patentamt*, Patentschrift Number 34749, Klasse 46, Tellier (August 15, 1885), Chapter 15.

Ackermann's quote is from A.S.E. Ackermann, "The Utilization of Solar Energy," *Smithsonian Institution Annual Report* (1915), p. 158. This article gives an excellent account of the history of sun motors.

94 ff. The primary source for the work of Willsie and Boyle comes from H.E. Willsie, "Experiments in the Development of Power from the Sun's Heat," *Engineering News*, Vol. 61, Number 19 (May 1909), pp. 511-514.

97 ff. Technical details on the sun power plants of Willsie and Boyle come from the following patents: U.S. Patent 1,101,000 (June 23, 1914); *U.S. Patent 1, 130,870* (March 9, 1915); *U.S. Patent 1,130,871* (March 9, 1915).

Chapter 9: The First Practical Solar Engine

101 The account of Shuman's discussion with his boys comes from a private interview with his son Frank Shuman.

Shuman's quote about the human race is from Frank Shuman, "Feasibility of Utilizing Power from the Sun," *Scientific American*, Vol. 110 (February 28, 1914), p. 179.

Professional journal quote about Shuman in "Power from the Sun's Heat," *Engineering News*, Vol. 61, Number 19 (May 13, 1909), p. 509.

The quote on the impracticality of using mirrors for sun machines is from Frank Shuman, "The Generation of Mechanical Power by the Absorption of the Sun's Rays," *Mechanical Engineering*, Vol. 33 (December 19, 1911), p. 1.

The quote about natural forces is from Frank Shuman, "Sun Power Plants Not Visionary," *Scientific American*, Vol. 110 (June 27, 1914), p. 60.

Shuman compared the fledgling solar industry with the aviation industry in Frank Shuman, "Solar Power," *Scientific American*, Vol. 71 (February 4, 1911), p. 78.

102 The *Engineering News* quote comes from "Power from the Sun's Heat," p. 509.

103 Shuman discussed the world potential for solar power in his article in *Mechanical Engineering,* p. 5.

104 *The Engineering News* quote about power is from "Power from the Sun's Heat," p. 509.

Shuman's comments about the need to keep the cost of sun plants low is from *Mechanical Engineering,* p. 2. This article details the annual operating costs of his sun plant.

Discussion of the finances of the Sun Power Company is taken from *Mechanical Engineering*, p. 6, as are the technical details of Shuman's second sun plant. Concerning the operation of the second Tacony plant see also A.S.E. Ackermann, "Shuman Sun-Heat Absorber," *Nature*, Vol. 89 (April 4, 1912), pp. 122-123; and Frank Shuman, "Power from Sunshine, A Pioneer Solar Power Plant," *Scientific American,* Vol. 105 (September 30, 1911), pp. 291-292.

105 For a thorough description of Shuman's special low-pressure engine design for the Tacony plant refer to R.C. Carpenter, "Tests of a Simple Engine," *Scientific American Supplement*, Vol. 74 (July 13, 1912), pp. 28-29.

107 ff. For a detailed discussion of the Egyptian sun plant see George Hally, "Sunpower—Its Commercial Utilization," *Institution of Engineers and Shipbuilders in Scotland*, Vol. 57 (April 21, 1914). Hally's article is the most complete single piece on Shuman's work. It describes all of his projects but centers on the Egyptian plant.

109 Information regarding the inaugural ceremonies of the Egyptian sun plant come from a private interview with Frank Shuman's son.

The economic analysis of the Egyptian sun machine is taken from "Sun Power Plant: A Comparative Estimate of The Cost of Power from Coal and Solar Radiation," *Scientific American Supplement*, Vol. 77, No. 1985 (January 17, 1914), p. 37.

The announcement that sun power was now practical is quoted from Frank Shuman, "Feasibility of Utilizing Power from the Sun," *Scientific American*, Vol. 110 (February 28, 1914), p. 179.

109 f. Details of Shuman's agreements with the Germans and British are from a private interview with Frank Shuman's son.

110 Shuman's quote on the vast potential for sun power is from Frank Shuman, "The Most Rational Source of Power: Tapping the Sun's Radiant Energy Directly," *Scientific American*, Vol. 109 (November 1, 1913), p. 350.

Chapter 10: Early Solar Water Heaters

115 Quote of early California homesteader is from an interview with Luba Perlin who lived on a ranch in the Soledad Mountains (California).

General information regarding domestic hot water systems at the turn of the century is from Royce G. Kloeffler, "Water Heating in the Home," *Kansas State Agricultural Bulletin 11* (June 1, 1921), pp. 1-35. This is probably the most thorough study of domestic hot water heating systems in that period. An excellent economic analysis of these units is included.

Theodore Hotchkiss, engineer and long-time resident of Monrovia, California, is quoted on the operation of hot water tanks attached to stoves.

115 f. Information regarding the operation of the side-arm water heater is from a private interview with William Ingalls, a retired plumber who opened up shop in 1916.

116 The price of artificial gas was taken from *Statistical Abstract of the United States* (Washington: Department of Labor, 1919), p. 576.

117 James Harrison, a retired Nevada prospector, gave the comment about bare-tank solar water heaters. Technical information regarding their performance can be obtained from F.A. Brooks, "Use of Solar Energy for Heating Water in California," *Report of the California Agricultural Experiment Station*, Bulletin 602 (1936), p. 7.

117 f. Size and price information, as well as advertising copy, for the Climax Solar Water Heater are taken from a sales brochure published in 1902 by the Clarence M. Kemp Manufacturing Company, Baltimore, Maryland. It was found in the Lawrence B. Romaine Trade Catalogue Collection, Department of Special collections, University of California at Santa Barbara.

120 The journalist is Lee McCrae, "Utilizing Sun Rays in a Home," *Countryside Magazine*, Vol. 23 (October, 1916), p. 203.

Information about the transfer of rights to the Climax Solar Water Heater was obtained from Patent Assignment Records, Customer Relations Office, U.S. Patent Office, Washington, D.C.

The figure of 1600 Climax Solar Water Heaters sold in southern California is taken from an advertisement in *The Los Angeles Times*, (June 3, 1900), Part 1, p. 7.

121 Information about Samuel Stratton is taken from "The Solar Water Heater," *Pasadena Daily Evening Star*, (August 17, 1896), p. 1.

121 f. The van Rossem quotes and other pertinent information are from a private interview with the former Pasadena resident.

122 Background information on Frank Walker appeared in James M. Guinn, *A History of California and Its Southern Counties* (Los Angeles: Historical Record Company, 1907), p. 871.

Technical information on Walker's solar water heater is from U.S. Patent 705,167 (April 19, 1898).

122 f. Concerning the ownership of the Solar Heating Company, makers of the Improved Climax, see *Articles of Incorporation for the Solar Motor Company* (California: July, 1903).

125 Installer Jim Bailey provided the comments about the performance of the Improved Climax in a private interview.

127 The J.J. Backus quote is from his article, "Praise for Solar Heater," *The Architect and*

Engineer of California, Vol. 8 (March, 1907), p. 98.

The performance of glass-covered tank solar water heaters is evaluated in Brooks, pp. 35-37.

Chapter 11: Hot Water—Day and Night

129 Information regarding the birth of the Day and Night solar water heater was obtained from private interviews with Edward "Ned" Arthur, one of the company's five original employees.

129 ff. Technical details on the Day and Night solar water heaters come from private interviews with Edward "Ned" Arthur, the late William Crandall, a Day and Night installer, and from the Day and Night Solar Water Heater Company Installation Guide (1912).

131 Crandall quotes on Day and Night installation from William Crandall interview.

The praise for the Day and Night solar heater is from "Solar Heater—A Monrovian Perfects a New and Improved Heater," *Monrovia Daily News* (July 3, 1909), p. 1.

132 Arthur quote about sales competition from Edward Arthur interview.

Day and Night's claim about hot water for laundry is from an advertisement appearing in *Monrovia Daily News* (July 17, 1909), p. 1.

Quote about the doubting rural Californians is from an interview with former schoolteacher Paul Squibb.

132 ff. The account of Bailey's solar water heaters freezing comes from an interview with Bailey's son, William J. Bailey, Jr.

135 Technical data concerning Bailey's freeze-protected solar water heater are from: Day and Night Solar Water Heater Company Sales Brochure (1914), U.S. Patent 1,242,511 (October 9, 1917); F.A. Brooks, "Use of Solar Energy for Heating Water in California," pp. 52-54.

136 The *Arizona Magazine* quote is from "Let the Sun Do the Work," *Arizona Magazine* (October, 1913), p. 9.

Arthur's comments about northern California sales are from private interviews.

139 The trade magazine observations come from "Old Sol Harnessed for Domestic Service," *Metal Worker, Plumber and Steam Fitter*, Vol. 81 (May 29, 1914), pp. 711-712. The number of solar heaters sold by Day and Night was estimated by examining corporate records in archives referred to as "The Morgue" at BDP (Bryant, Day and Night, and Payne) Corporation located in La Puente, California. A Day and Night sales brochure, "The Day and Night Solar Water Heater" (published circa 1925) confirms these estimates.

140 Comments about the sales strategy of the gas companies are from an interview with Robert Carroll, an early Santa Barbara plumber.

141 William J. Bailey's comments are from a private interview.

Chapter 12: A Flourishing Solar Industry

143 Background information on the Miami land boom is from Fredrick Lewis Allen, *Only Yesterday* (New York: Harper and Brothers, 1931), pp. 270-289.

The price of electricity and artificial gas in Florida is taken from *Statistical Abstract of the United States* (Washington: Department of Labor, 1923), pp. 586-587. Prices taken from Jacksonville have been adjusted for Miami.

The information about Bailey's deal with Carruthers comes from an interview with William J. Bailey, Jr.

143 ff.   The Heath quotes are from an interview with Harold Heath, former sales manager of The Solar Water Heater Company. Because of Heath's familiarity with the history of Miami and The Solar Water Heater Company, he is one of the principal sources for this chapter.

144   The quote about the bungalow is from a real estate advertisement for the L.J. Ursem Company, *Miami Herald* (April 4, 1926), p. 12. During the 1920's numerous advertisements such as this appeared in Miami's newspapers.

Munroe's statements are from an interview with William D. Munroe, an early Miami resident.

Information about the importance of The Solar Water Heater Company in the Miami construction industry is from "Firms Keep Pace with the Rapid Growth of Miami," *Miami Herald* (April 9, 1925), p. 1-G.

144 ff.   Information regarding Ewald's reopening of the Solar Water Heater Company is from an interview with Harold Heath.

146   Regarding The Solar Water Heater Company's use of soft copper in their collectors see Charles F. Ewald, *Open Letter* (July 30, 1935).

146 f.   Many details about Ewald's improvements come from interviews with Heath and Walter Morrow, currently with The Solar Water Heater Company.

147   Technical information on the Duplex has been obtained from U.S. Patent 2,065,653 (December 29, 1936); apparently because Carruthers still retained ownership of The Solar Water Heater Company, this patent was issued in his name. See also Harold M. Hawkins, "Domestic Solar Water Heating in Florida," *Florida Engineering & Industrial Experimental Station*, Bulletin 18 (September, 1947). Hawkin's pamphlet provides a thorough discussion of the design features of the Duplex as well as other Florida solar water heaters.

148   Heath's quote about sizing a system is from interviews with Harold Heath.

149 ff.   Information about the F.H.A. loan program is from interviews with Harold Heath.

151   Ewald's quote is taken from "Solar Water Heater Demand is Increased," *Miami Herald* (August 5, 1935), p. 7.

151 f.   Concerning the highly competitive years from 1937-1941 see Jerome E. Scott, *Solar Water Heater Industry in South Florida, 1923-1974* (Washington: National Science Foundation, 1975), pp. 3-30. Scott's work is a comprehensive historical study of the use of solar water heaters in southern Florida.

152   Statistics on the number of Miamians using solar water heaters are from Scott, p. 49, as well as from numerous articles of the period.

Facts about large solar water heater installations as well as those used in housing projects come from "Let Sol Do the Work," *Domestic Engineering*, Vol. 157 (May, 1941), pp. 48-49 and pp. 124-127.

Concerning interest of public housing officials in solar water heating see *Public Housing Design* (Washington: National Housing Agency, Federal Public Housing Authority, June, 1946), pp. 234-235. The information on solar water heater installations in the Caribbean and Central America came from the files of the Solar Water Heater Company of Miami.

155 The engineering article that warned of the danger of using dissimilar metals is Harold L. Alt, "Sun Effect and the Design of Solar Heaters," *Transactions of the American Society of Heating and Ventilating Engineers (A.S.H.V.E.)*, Vol. 41 (1935), pp. 131-148.

The economic problems that the solar industry faced after World War II are discussed in Scott.

Chapter 13: Solar Communities of Europe

159 The quote by Magalhaes is from Gabriel Magalhaes, *The History of China* (London: Thomas Hewborough, 1688), pp. 266-271. Also, on the continuing tradition of solar architecture and town planning in China, see Andrew Boyd, *Chinese Architecture and Town Planning* (Chicago: University of Chicago Press), pp. 49-121; D.G. Mirams, *A Brief History of Chinese Architecture* (Shanghai: Kelly and Walsh Limited, 1940), pp. 24-30; and Nelson Wu, *Chinese and Indian Architecture* (New York: Braziller, 1963), p. 35.

For continuation of solar architecture among the Greeks and Turks see Theodore Weigand and Hans Schrader, *Priene* (Berlin: G. Reimer, 1904), p. 290.

Palladio called Vitruvius his master in Andrea Palladio, *The Four Books of Architecture* (London: Issac Ware, 1738), "Preface to The Reader." Concerning his prescriptions for the orientation of winter and summer rooms see Palladio, 2.2.

159 f. Repton's comments about the misapplication of Greek principles of solar architecture are from Humphrey Repton, *Observations on the Theory and Practice of Landscape Gardening* (London: T. Bensley, 1803), p. 196, footnote 1. See also p. 195.

160 Information on the European slums comes from Henry R. Aldridge, *The Case for Town Planning* (London: The National Town Planning Council, 1923). Aldridge's book contains an exhaustive study of urban planning and housing in Europe, up to the beginning of the twentieth century.

The physician's quote is from Augustin Rollier, M.D., *Heliotherapy* (London: Oxford University Press, 1913), p. 149.

The Parliament legislation is paraphrased in Charles M. Robinson, *City Planning* (London: A. Rivers, Ltd., 1909), p. 236. The German building code is taken from Robinson, p. 236.

162 Information on Port Sunlight is from Walter L. George, *Labour and Housing at Port Sunlight* (New York: G.P. Putnam's Sons, 1916), p. 23.

For Unwin's work see Raymond Unwin, *Town Planning in Practice* (London: T. Fisher Unwin, 1909). Unwin's comments are found on p. 23.

163 f. For Rey's work and comments, refer to Augustin Rey, "La Ville Salubre de l'Avenir," *International Congress of Local Authorities*, Vol. 1 (Ghent, 1913), pp. 217-224.

164 Rey's research stimulated Garnier's interest in solar architecture. Information on his work comes from Tony Garnier, *Une Cité Industrielle: Étude Pour la Construction des Villes*

(Paris: A. Vincent, 1918), and Christophe Pawlowski, *Tony Garnier: Et les Debuts de l'Urbanisme Fonctionel en France* (Paris: Centre de Recherche d' Urbanisme, 1967), p. 240.

165 ff. On the development of "modern" architecture in Germany after World War I, see Lewis Mumford, "Machines for Living," *Fortune,* Vol. 7 (February, 1933), pp. 82-88; Catherine Bauer, *Modern Housing* (Boston: Houghton Mifflin Company, 1934). Bauer's book is probably the most authoritative source on the housing movement of the 1930's. See also Henry Wright, *Rehousing Urban America* (New York: Columbia University Press, 1935). Wright's book discusses the evolution of planning communities toward the sun in both Germany and other European countries during the 1920's and 1930's.

166 f. Breuer's comments are from a private interview. He was one of the initial members of Germany's influential design school, the Bauhaus.

167 f. Gropius is quoted from Walter Gropius, *Scope of Total Architecture* (New York: Collier Books, 1962), p. 110. Gropius, the founder of the Bauhaus, was instrumental in promoting interest in orienting buildings to take advantage of the sun.

167 f. Information on Siemenstadt is from D. Rentschler and W. Schirmer, *Berlin und Seine Bauten* (Berlin: Von Wilhelm Ernst und Sohn, 1974), pp. 398-406.

168 Mumford's comments are from Lewis Mumford, p. 87. He later lost his enthusiasm for east/west oriented buildings and opted for a southern orientation. Von Moltke's comments are from a private interview. For the empirical studies that showed why the *Zeilenbau* plan did not work as planned, see Paul Schmitt, "Die Besonnongsverhaltnisse an Stadtsgrassen und Die Gustigste Blockstellung," *Bauwesen,* Vol. 80 (1930), especially p. 109; and Ludwig Von Hilberseimer, "Raumdurchsonnumg," *Moderne Bauformen,* Vol. 34, Part 1 (January, 1935), pp. 29-36. Schmitt's study was primarily concerned with the orientation of buildings relative to the insolation of streets. Von Hilberseimer performed a thorough investigation of sunlight penetration of buildings at the solstices and equinoxes.

168 f. Concerning solar-oriented single-family residences, refer to Hugo Haring, "Bemerkingen zum Flachbau," *Moderne Bauformen*, Vol. 33, Part 2 (November, 1934), pp. 619-627.

169 Haring's quote is from Hugo Haring, p. 620.

169 f. Although the Nazis sought to re-establish "traditional" architecture, there was no clear-cut definition of what this architecture would look like. See Barbara M. Lane, *Architecture and Politics in Germany* (Cambridge: Harvard University Press, 1968), p. 185. Nazi opposition to "modern" architecture was clear; see Bauer, p. 222. Hitler's fear of a concentrated proletariat was based on the 1919 German revolt and Austrian socialist uprisings in which workers "shot from the windows" of their communal apartments and barricaded themselves inside "as in so many fortresses." See Leonardo Benevolo, *History of Modern Architecture* (Cambridge: M.I.T. Press, 1971), p. 526.

170 On solar-oriented housing in other European countries see Catherine Bauer, Part 4 and Appendix. Examples of solar architecture in Holland are to be found in Jacobus J.P. Oud, *Hollandische Architektur* (Munchen, 1926). The Swedish workers' village of Kvarnholmen was inspired by Garnier's Cité Industrielle. See Pawlowski, p. 221.

170 f. On Neubuhl refer to Alfred Roth, *Die Neue Architecktur* (Zurich: Les Editions d'Architecture, 1948), pp. 71-90. Roth's brother Emil was one of the seven architects who designed this solar community.

Authors' note: The terms "solar city" or "solar community" were not used during the period described in this chapter.

Chapter 14: Solar Heating in Early America

173 Background information on Acoma is from Mrs. William T. Sedgwick, *Acoma, The Sky City* (Cambridge: Harvard University Press, 1926). Technical analysis of the solar heat retention of dwellings at Acoma comes from Ralph Knowles, *Energy and Form* (Cambridge: M.I.T. Press, 1974), pp. 28-33. Although there are no written records indicating that the Pueblo Indians built with the sun in mind, Knowles concludes from his research that they had a remarkably sophisticated knowledge of solar orientation.

On Spanish Colonial architecture in America see Donald R. Hannaford and Revel Edwards, *Spanish Colonial or Adobe Architecture of California: 1800-1850* (New York: Architectural Book Publishing Company, Inc., 1931), p. 1.

The Breckenridge quote appeared in Thomas Breckenridge, *Modern Chivalry* (Philadelphia: Petersen, 1856). Specific design details on the salt-box house comes from "Two Craftsman Houses: A Plaster Dwelling That Is Suitable for Either Town or Country," *The Craftsman*, Vol. 15 (January, 1909) pp. 472-480. On American traditional architecture see Lewis Mumford, *Roots of Contemporary Architecture* (New York: Reinhold, 1952), pp. 11-12.

176 Concerning Price's architectural designs see Bruce Price, *A Large Country House* (New York: William T. Comstock, 1887).

176 ff. Information on Atkinson is from William Atkinson, "The Orientation of Buildings and of Streets in Relation to Sunlight," *Technology Quarterly*, Vol. 18 (September, 1905), pp. 204-206; and William Atkinson, *The Orientation of Buildings or Planning for Sunlight* (New York: John Wiley & Sons, 1912). His quote on skyscrapers appeared in *Technology Quarterly*, p. 204. The legislation that Atkinson helped enact (House Bill No. 775, Commonwealth of Massachusetts, January 26, 1905) limited building height to 2½ times the width of the streets adjacent to them. Details on Atkinson's "Sun-Box" and "Sun-House" are from *The Orientation of Buildings or Planning for Sunlights*, pp. 62-78.

Chapter 15: An American Revival

181 Information on the R.I.B.A. study is from "The Orientation of Buildings," *Journal of the Royal Institute of British Architects* (September 10, 1932), pp. 777-797. The report was primarily concerned with sunlight penetration of buildings and only touched on the heating aspect of solar orientation.

On Henry Wright refer to "News of The Month," *Architectural Forum*, Vol. 79 (April, 1936), p. 248. Information was also obtained from an interview with his son, Henry (N.) Wright.

About the A.S.H.V.E. study see W.W. Shaver, "Heat-Absorbing Glass Windows," *Transactions of the American Society of Heating and Ventilating Engineers*, Vol. 41, No. 1018 (1935), p. 297.

181 ff. Information regarding Henry (N.) Wright's study of orientation and solar heat comes from a personal interview. Although the study was not published by its original sponsor, Wright published parts of it in "Products and Practice," *Architectural Forum*, Vol. 68 (June, 1938), pp. 18-22. Although his figures were somewhat exaggerated, this report helped to stimulate further interest for solar orientation in America.

183 ff. The Keck story and comments are from extensive interviews with the architect. More than

any one else, George Fred Keck promoted the development of solar architecture in America during the late 1930's and early 1940's.

184   Sloan's comment comes from "Two Developments in Glenview, Ill.," *Architectural Forum*, Vol. 80 (March, 1944), p. 86.

185   The *Business Week* quote is taken from "Sun-Heated House," *Business Week* (October 5, 1940), p. 20.

The story of Solar Park comes from "Solar Houses," *Business Week* (December 27, 1941), p. 42; "Two Developments," p. 86; and from an interview with Sloan's wife, Katherine.

186 ff. Concerning the Libbey-Owens-Ford study see *Solar Heating for Post-War Dwellings* (Toledo: Libbey-Owens-Ford Glass Company, 1943), and "Solar Heating," *Architectural Forum*, Vol. 79 (August, 1943), pp. 6-8, and 114.

187   The engineers conclusions are taken from "Solar Heating," p. 8.

The conservative statement by Knopf and Peebles is from *Solar Heating for Post War Dwellings*, p. 7, as are the comments on total illumination.

189   Carsten's letter is quoted from *The Solar Home* (Rockford, Illinois: Green's Ready-Built Homes, 1944), p. 9.

The Bennetts' home is reported in Ralph Wallace, "Sun Is the Fireman in New Glass Homes," *Baltimore Sun* (December 12, 1943), Section 1, p. 1. Wallace lists and discusses a number of "solar" homes in this article.

The *Newsweek* quote is from "Dream Houses—U.S. and British," *Newsweek* (September 6, 1943), p. 103.

The first *House Beautiful* quote comes from "Can an Old House Be Remodeled for Solar Heating?," *House Beautiful*, Vol. 87 (June, 1945), p. 75. The following quote is from "Did You Know That the Sun Can Heat Your Home in Winter?," *House Beautiful*, Vol. 85 (September, 1943), p. 63. Elizabeth Gordon, then editor of the magazine, often featured articles on solar design during the 1940's and early 1950's.

Ladies Home Journal's modern architecture project is found in *Tomorrow's Small House* (New York: The Museum of Modern Art, 1945), p. 4.

190   Postwar housing statistics are taken from *The Handbook of Basic Economic Statistics*, Vol. 32 (January, 1978), p. 194.

Information concerning the prefabricated solar homes is from: "The Solaray House," (Boston: Solaray Corporation, 1947); a private interview with Peter Morton, son of Oliver P. Morton, one of the architects of the Solaray Home; "Wickes, Inc.," *Architectural Forum*, Vol. 90 (January, 1949), p. 98; an interview with one of Wickes' architects, Orin Thomas; and an interview with Edward Green, the founder of Green's Ready Built Homes.

191   Sloan's Northbrook development is discussed in "It's Here—Solar Heating for Post-War Homes," *American Builder*, Vol. 65 (September, 1943), pp. 34-36 and p. 94. It is very difficult to document how many solar developments went up but Sloan himself built at least four.

191 f. The information and comments on Hutchinson's study are from F.W. Hutchinson, "Solar House: A Full-Scale Experimental Study," *Heating and Ventilating*, Vol. 42 (September, 1945), pp. 96-97; F.W. Hutchinson, "Solar House; Research Progress Report," *Heating and*

Ventilating, Vol. 43 (March, 1946), pp. 53-57; and F.W. Hutchinson, "Second Research Progress Report," Vol. 44 (March, 1947), pp. 55-59. Hutchinson's study of solar heat retention by large south-facing glass windows proved to be inconclusive as did the first L.O.F. study. The equation he developed to help architects determine how much glass to use on a south wall for optimum solar heat retention can be found in F.W. Hutchinson and W.P. Chapman, "A Rational Basis for Solar Heating Analysis," *Heating, Piping, and Air Conditioning,* Vol. 18 (July, 1946), pp. 109-117.

193 f.   Information on Brown's work is from an interview with the Tucson architect and from "A Black Wall Stores Winter Sun Heat," *House Beautiful,* Vol. 92 (April, 1950), p. 150; see also "They Heat Their House with Sunshine," *Better Homes and Gardens,* Vol. 31 (February, 1953), p. 175 and p. 199.

194   Keck is quoted from his paper on retractable awnings in *Program for a Course Symposium: Space Heating with Solar Energy* (Cambridge: M.I.T., August 21-26, 1950).

195   The extra costs involved in solar developments is from an interview with Frank Burns, who, in cooperation with Howard Sloan, built a large solar development in Colorado Springs, Colorado.

Information on the building code problems which prefabricated home-builders faced comes from interviews with George Fred Keck, Edward Green, Orin Thomas, and Henry (N.) Wright.

A good selection of typical solar homes built during this period appear in M.D. Gillis, *McCall's Book of Modern Houses* (New York: Simon and Schuster, 1951).

Chapter 16: Solar Collectors for House Heating

197 ff.   For a detailed description of Morse's work see Edward S. Morse, "The Utilization of The Sun's Rays in Heating and Ventilating Apartments," *The Proceedings of The Society of Arts of The Massachusetts Institute of Technology* (1885).

197   *Scientific American* comment came from "Heating by Sunshine," *Scientific American,* Vol. 46 (May 13, 1882), p. 288.

198   Concerning newspaper accounts of Morse's solar heater see the *Troy Daily Times* (January 19, 1893) and *The Salem Gazette* (January 14, 1893).

200   The letter from the Climax Solar Water Heater Company to Morse came from Charles A. Saxton, Los Angeles Agent (September 11, 1897).

Morse's rejoinder to the absurd newspaper stories appeared in the *Salem Observer* (January 28, 1893).

For Morse's comment about the usefulness of his invention in fuel-short areas, see the *Salem Observer* (January 28, 1893).

The comments about Godfrey Lowell Cabot are from "Solar Attack," *Time,* Vol. 31 (June 6, 1938), p. 20.

For the goals of the M.I.T. solar project, see Hoyt Hottel, "Converters of Solar Energy," *Annual Report of the Smithsonian Institution* (1941), pp. 152-156; and H.C. Hottel and B.B. Woertz, "The Performance of Flat-Plate Solar Heat Collectors," *American Society of Mechanical Engineers Transactions* (February, 1943), p. 91.

200 ff. Information about the first M.I.T. solar house comes from Hottel and Woertz, pp. 91-104, and from personal conversations with the two of them. The equation they developed concerning the performance of the flat-plate collector is discussed in Hottel and Woertz, pp. 93-103.

203 ff. The story of Löf's solar heating system was obtained from interviews with Dr. George Löf; see also the Engineering Experiment Station, University of Colorado, "Solar Energy Utilization for House Heating," (Washington: Office of Production, Research, and Development; War Production Board, May 18, 1946), Project #459, Contract #100.

206 ff. Data on radiant water walls, M.I.T.'s second solar heating experiment, comes from interviews with Albert G.H. Dietz, Professor of Civil Engineering, M.I.T., and Hoyt Hottel, head of the M.I.T. solar research project. See also Albert G.H. Dietz and Edmund L. Czapek, "Solar Heating of Houses by Vertical South Wall Storage Panels," *Heating, Piping, and Air Conditioning,* Vol. 22 (March, 1950), pp. 118-125.

208 ff. The details of M.I.T.'s third solar experiment came from interviews with Mr. Dietz and Mr. Hottel; see also *Solar House Heating* (Cambridge: M.I.T. Solar Energy Conversion Research Project, December, 1952). Specific data on performance is from the winter of 1951-1952. Although test homes investigated by F.W. Hutchinson and Knopf and Peebles showed little heat gain due to south-facing windows, this lived-in home showed very different results.

211 f. Maria Telkes discussed solar heat storage and her ideas concerning salts of fusion in Maria Telkes, "Solar House Heating—A Problem of Heat Storage," *Heating and Ventilating,* Vol. 44 (May, 1947), pp. 71-73.

212 Albert Dietz's comments are from a private interview.

213 f. A detailed description of the Dover House can be found in "Test House Heated Only by Solar Heat," *Architectural Record,* Vol. 105 (March, 1949), pp. 136-137.

214 Esther Nemethy's dissatisfaction with life in the Dover House was expressed in a private interview. Her husband provided more detailed information on the system performance, which he monitored for almost three years.

214 ff. Information regarding the Rose Elementary School was obtained from interviews with Arthur Brown, the designer.

216 Details of the fourth M.I.T. solar house come from C.D. Engbretson and N.G. Ashar, "Progress in Space Heating with Solar Energy," Report No. 60-WA-88, The American Society of Mechanical Engineers (July 28, 1960). Interviews with Hottel and Dietz were also helpful.

Chapter 17: Postwar Energy Perspectives

221 Oil and natural gas prices are from *Statistical Abstract of the United States* (Washington: Department of Labor, 1965), p. 366 and p. 369. In 12 out of 15 major metropolitan areas in the U.S.A., one thousand cubic feet of natural gas sold for under $1.00 if the customer used at least 100 therms.

Data on crude oil reserves come from *Statistical Abstract of the United States* (1955 and 1965), p. 750 and p. 721, respectively. For data on natural gas reserves see *Statistical Abstract of the United States* (1965), p. 723.

The gas company executive quoted anonomously currently works for the Southern California Gas Company.

The electric company executive quoted is currently with Southern California Edison. He also wished to remain anonymous.

A graphic example of this preferential rate structure is presented in *Statistical Abstract of the United States* (1965), p. 369.

222   Data on electricity production are from *Statistical Abstract of the United States* (1969), p. 511. For the data on natural gas production, see *Statistical Abstract of the United States* (1965), p. 672. On U.S. fuel consumption as a whole, see *Historical Statistics of the United States* (Washington: Department of Commerce, 1975), Part 2, p. 819.

Eric Hodgins is quoted from his article, "Power from the Sun," *Fortune,* Vol. 148 (September, 1953), p. 130.

222 f.   Scarlott's comments are from a private interview. With Eugene Ayres he wrote *Energy Sources: The Wealth of the World* (New York: McGraw-Hill), 1952.

223   Remarks by the President's Materials Commission quote in "The Promise of Technology," *Resources for Freedom* (Washington: President's Materials Commission, June, 1952), p. 220 and p. 217.

Funding for research and development of solar energy was almost nil throughout the Postwar years. See S. David Freeman, "Is There An Energy Crisis: An Overview," *The Annals of The American Academy,* Vol. 410 (November, 1973).

Industry's stake in nuclear power is outlined in "How To Make Money Out of The Atom," *U.S. News and World Report,* Vol. 37, (August 13, 1954), pp. 24-25.

Alvin Weinberg wrote about this international conference in "Nuclear Journey Through Europe," *Bulletin of the Atomic Scientists,* Vol. 10 (June, 1954), pp. 215-217.

223 f.   Concerning Europe's fuel problems in the 1950's and their choice of nuclear power, see "Fuel Shortage Sparks West Europe's Atomic Energy Plans," *Business Week,* Vol. (December 29, 1956), pp. 76-78. Quote appeared on p. 76.

225   The report on Eisenhower's commencement of America's first atomic power plant is from "A Wand Wave, A New Era," *Life,* Vol. 37 (September 20, 1954), pp. 141-142.

226   Farber and Reed's warning about nuclear power can be found in E.A. Farber and J.C. Reed, "Practical Applications of Solar Energy," *Engineering Progress at The University of Florida,* Leaflet Series, Vol. 10, No. 2 (November, 1956), p. 1.

Russler's review of atomic and solar energy can be found in George W. Russler, "Nuclear or Solar Energy: Which Is More Practical for Space Heating?," *Heating, Piping, and Air Conditioning,* Vol. 31 (February, 1959), pp. 106-109. The quotes used are from pp. 107, 109, and 109, respectively.

227   Becquerel observed that when the sun struck one of two electrodes immersed in a conductive fluid, an electrical current was produced. Concerning Becquerel's pioneering research into the photovoltaic effect see Edmund Becquerel, "Memoire sur Les Effets Electriques Produits sous L' Influence es Rayons Solaires," *Comptes Rendu de l' Academie des Sciences,* Vol. 9 (1839), pp. 561-565.

Concerning Fritts' work see C.E. Fritts, "On The Fritts' Selenium Cells And Batteries," *Van Nostrand's Engineering Magazine,* Vol. 32 (1885), pp. 388-395. For another evaluation of this work see Werner Siemens, "On The Electromotive Action of Illuminated Selenium Discovered by Mr. Fritts of New York," *Van Nostrand's Engineering Magazine,* Vol. 32 (1885), pp. 514-516.

228 With the rediscovery of selenium solar cells in the 1930's, one journalist wrote, for example, "If this dream of power from the sun is realized, it will go far to make industry independent of the fast dwindling coal supply of the world." See "Magic Plates Tap Sun for Power," *Popular Science Monthly,* Vol. 118 (June, 1931), p. 134.

228 ff. The details of Chapin, Fuller, and Pearson's work were derived from conversations with Chapin and Pearson. See also D.M. Chapin, "The Direct Conversion of Solar Energy to Electrical Energy," in A.M. Zarem and D.D. Erdway, *Introduction to the Utilization of Solar Energy* (New York: McGraw-Hill, 1963), pp. 153-189.

229 The futuristic description of what solar cells would do appeared in "Solar Cell Is Ready for Commercial Jobs," *Business Week* (July 20, 1957), p. 6.

233 The *Times* editorial comment concerning the need for a more even-handed policy concerning solar development appeared in the *New York Times* (April 24, 1954), p. 24.

Chapter 18: Worldwide Solar Water Heating

235 Gonen Yissar, son of Levi Yissar, is quoted from private correspondence.

Levi Yissar's comments are from "Do You Have Hot Water?"*Davar HaShavuah* (February 7, 1952), p. 6. This article appears only in Hebrew as do the following articles from *Marriv.*

Details of Yissar's entry into the commercial field come from "The Commercialization of Solar Water Heaters Begins," *Maariv* (August 19, 1953).

235 f. Regarding Ben Gurion's solar heater see "Mrs. Ben Gurion Becomes Acquainted with Her House and Neighbors," *Maariv* (December 14, 1953).

236 Gonen Yissar's claims about some of his father's colleagues come from private correspondence. Even though others did enter the field, Yissar's company was still the major solar water-heater manufacturer in Israel at this time.

237 Information on Miromit's entry into the solar industry is from private correspondence with Dr.Harry Tabor. His comments are quoted from that correspondence.

Concerning details of Miromit's improvements on the solar water heater, see John I. Yellot and Rainier Sabotka, "An Investigation of Solar Water Heater Performance," *American Society of Heating, Refrigeration, and Air Conditioning Engineers (A.S.H.R.A.E.) 71st Annual Meeting,* paper (June 29-July 1, 1964). See also Rainier Sabotka, "Miromit Solar Water Heater," *Sun At Work,* Vol. 5 (Fourth Quarter, 1965), pp. 12-13.

238 Sabotka's quote concerning Miromit's export business is from Rainier Sabotka, "Solar Water Heaters from Katmandu to Bamako," *Sun at Work,* Vol. 10 (Second Quarter, 1966), p. 10. Although Miromit exported a large number of solar water heaters, their primary market remained in Israel.

238 f. Information about the early development of solar water heaters in Australia and Morse's role in this development is from personal correspondence with Roger Morse and F.G. Hogg, former Secretary-Treasurer of the International Solar Energy Society.

Information on C.S.I.R.O.'s aid to solar water-heater manufacturers came from an interview with Mr. Keith Jenkins of the Solarhart Company.

239   Regarding the placement of solar water heaters on government buildings in tropical areas, refer to Roger N. Morse, "Solar Energy Research: Some Australian Investigations," *C.S.I.R.O.* (October, 1959), pp. 26-27.

Morse's comment about Australian government policy is from Roger N. Morse, "Solar Energy in Australia," *Ambio. Royal Swedish Academy of Sciences* (1977), p. 213.

Jenkins' comments are from a private interview.

Data on the number of flat-plate collectors sold in Australia are taken from correspondence with the Manufacturing Division, Australian Bureau of Statistics, Canberra, Australia.

240   Information about Mr. Rome's firm is from private correspondence with the Johannesburg engineer.

Regarding Austin Whillier's influence on Rome see Austin Whillier "The Prospects for Engineering Utilization of Solar Energy," *The South African Mechanical Engineer,* Vol. 5 (October, 1956), pp. 85-88.

The quote about solar heating and smog control is made by Mr. R.G. Collender in Austin Whillier, p. 88.

240 f.   Information on the changing freight rates and their effect on solar water heaters in South Africa was obtained from private correspondence with M.D. Lewis Rome.

241 ff.   Information on the development of the solar water-heater industry in Japan comes largely from private correspondence with Professor Ichimatsu Tanashita.

242   Regarding technical data on the basin-type solar water heater, see Ichimatsu Tanashita, "Present Status of Solar Water Heaters in Japan," *Transactions on the Use of Solar Energy: The Scientific Basis,* Vol. 3, Part 2 (October 31-November 1, 1955), pp. 67-68. This paper also discusses the history of solar water heaters in Japan from the 1940's through the mid-1950's.

242 f.   Information on the Kaneko-Kogyosho Company is from correspondence with Professor Tanashita.

243 ff.   Concerning vinyl solar water heaters, refer to Ichimatsu Tanashita, "Recent Development of Solar Water Heaters in Japan," *Proceedings of the United Nations Conference on New Sources of Energy, Rome, 1961* (New York: United Nations, 1964), pp. 104-105; and a sales brochure, Taiyosha & Company Ltd. (1960). Taiyosha was a major manufacturer of vinyl solar water heaters during this period.

245   Statistics on the sales of vinyl solar heaters are from Ichimatsu Tanashita, "Present Situation of Commercial Solar Water Heaters in Japan," *The International Solar Energy Society Conference, Australia* (1970).

245 f.   For detailed information on tank-type solar water heaters, see Ichimatsu Tanashita, "Recent Development of Solar Water Heaters in Japan," pp. 105-106 and "Present Situation of Commercial Solar Water Heaters in Japan," pp. 2-4. These papers include the sales figures quoted.

247   Information on the energy situation in Japan during the 1960's is from correspondence with Professor Tanashita.

This index covers Chapters 1 through 19; the Foreword and Notes are not included. Page numbers printed in **boldface** indicate an illustration or photograph.

Index

The following individuals and institutions provided maps, illustrations, photographs or other artwork for this book. The page number and position (*t*-top, *b*-bottom, *l*-left, *r*-right and *m*-middle) of each item on the page are listed before the source.

iii *mr*: Swissair Photos, Zürich; xii, 1: Adapted from a seventeenth-century woodcut; 2, 4, 16*b*, 18*t*, 21, 24*t*, 26, 50*r*, 56, 96, 111, 127*l*, 133*br*, 163, 170*t*, 174*t*, 242*r*, 245*r*: Sara Boore; 6*r*, 7*t*: From *Excavations at Olynthus*, by D. M. Robinson and J.W. Graham, ©1938, 1946 by Johns Hopkins University Press; 7*b*: Courtesy of The Royal Ontario Museum, Toronto, Canada; 9*r*: Reprinted from *Enciclopedia dell'Arte Antica*, ©1965 by Instituto della Enciclopedia Italiana Fondata da Giovanni Treccani; 10, 17, 18*b*, 22*l*, 24, 25: Norman Neuerburg; 18*t*, 23*b*: Adapted from drawings by Edwin D. Thatcher; 35, 36, 49*b*: Special Collections, University Research Library, University of California, Los Angeles; 39: The Bettmann Archive, Inc.; 44-45, 46*l*, 46*r*: William Andrews Clark Memorial Library, University of California, Los Angeles; 57: Master and Fellows of St. John's College, Cambridge, England; 58, 79: Smithsonian Institution Archives; 60, 61: Adapted from a 1910 stock certificate of The Sun Power Company; 87, 125*t*, 125*b*, 126, 127*r*, 172: Los Angeles County Museum of Natural History; 89: Department of Architecture, Arizona State University; 92-93: Bibliothèque National, Paris; 100, 102*l*, 102*r*, 103*b*, 105*t*, 105*b*, 106, 107*l*, 107*r*, 108*t*, 108*bl*, 108*br*, 110*t*, 110*b*: Frank R. Shuman; 112, 113: Adapted by Edward Wong-Ligda from an advertisement of the Solar Water Heater Company; 114, 119*t*, 119*bl*, 119*br*: Special Collections, Romaine Collection, University of California, Santa Barbara; 118*l*: Alice Kemp Travers; 118*r*: The C. M. Kemp Manufacturing Company; 120: *Los Angeles Times*; 121*t*: Title Insurance and Trust Company; 121*b*, 133*t*, 140: Historical Collections, Security Pacific National Bank; 130*r*, 134*b*: William J. Bailey, Jr.; 137: Reprinted from *Popular Mechanics*, ©1935 by The Hearst Corporation; 154*l*, 154*r*: William D. Munroe; 156, 157: ©1980 by Edward Wong-Ligda, adapted from a 1930's Cuban solar water heater company trademark; 158: Akademie der Kunste, Arthur Köster; 165, 167: Wilhelm Ernst und Sohn Verlag, Berlin; 166: Gehag Archiv, Berlin; 171*t*: Dr. Alfred Roth; 171*b*: Swissair Photos, Zürich; 174*b*: Dick Kent; 175*b*: Courtesy of The Southwest Museum; 180, 182*r*, 184, 187*b*, 188*b*: Photographs by Hedrich-Blessing; 182*tl*: Henry (N.) Wright; 182*tr*: George Fred Keck; 185*l*: Katherine Sloan; 185*r*: Libbey-Owens-Ford Company; 186: William Keck; 187*t*: Reprinted from *Architectural Forum*, ©1943 by Whitney Communications Corporation; 188*t*: Jane Heron; 190, 204*b*: Reprinted from *Popular Science* with permission, ©1945, 1947 by Popular Science Publishing Company; 191: Photograph by Gottscho-Schleisner; 192*tl*: Photograph by Stoody; 192*b*, 193, 194: Maynard Parker; 196, 201, 202, 206, 207, 209, 210*t*, 210*b*: M.I.T. Historical Collections; 198*r*, 199*tr*, 199*b*: Peabody Museum of Salem; 204*t*, 205: Dr. George O. G. Löf; 208: Reprinted from *Heating, Piping and Air Conditioning*, ©1950 by Reinhold Division of Penton/I.P.C. Publishing Company; 212: Eleanor Raymond; 213: Wide World Photos; 215*b*: Ray Manley; 217: Michael Vaccaro; 218, 219: ©1980 by Linda Goodman; 220, 232: National Aeronautics and Space Administration; 222*l*, 225: Edison Electric Institute, Washington, D.C.; 222*r*: Southern California Gas Co.; 224*l*: Reprinted from *U.S. News and World Report*, ©1954 by U.S. News and World Report, Inc.; 224*r*: Photograph by Walter Sanders, from *Life*, ©1954 by Time, Inc.; 227: LaRousse, Paris; 228, 230*t*, 230*b*: Bell Telephone Labs; 231: International Rectifier Company; 234, 250*b*: Miromit Company, Tel Aviv; 236*tl*, 236*tr*: Gonen Yissar; 237: John I. Yellott; 238, 239: R. N. Morse; 240: Solahart Company, Perth; 241: M. D. Lewis Rome; 243, 244, 245*l*, 246: Ichimatsu Tanashita; 248: Solaron Corporation; 250*t*: Azuma-Koki Company, Tokyo; 251: Photograph by Chad Ankele; 252: Acurex Corporation, Mountain View, California.

Credits

In addition to the people mentioned on page *v*, the authors would like to thank those who generously gave advice, criticism, personal accounts, photographs or access to their files. To these people and to all others who helped us, we offer our deep gratitude: Elliot L. Beasley, Marcel Breuer, Frank Burns, Dr. Chi-Yun Chen, Peter Clark, Sharon Cosner, Elizabeth Gordon, Edward Green, James Harrison, Dr. F. G. Hogg, Dr. Hoyt Hottel, Dr. F. W. Hutchinson, Keith Jenkins, William Keck, A.I.A., William Kemp Lehman, Dr. John McCabe, Walter Morrow, Roger N. Morse, Peter Morton, Lewis Mumford, William D. Munroe, Dr. Anthony Nemethy, Esther Nemethy, Gordon Pearson, W. Robbins, Jr., Dr. Alfred Roth, M. D. Lewis Rome, P.E., Katherine Sloan, Dr. Harry Tabor, Dr. Maria Telkes, Edwin D. Thatcher, A.I.A., Orin Thomas, A.I.A., Alice Kemp Travers, and Dr. Byron B. Woertz.

The following museum curators and librarians also aided our search: Hilda Bohem, Chris Braun, George Mackovec, Bill Massa, Victor Plukas, Vicki Steele, and Corliss Wendling.

Translations of primary source material from Greek, Latin, Hebrew, French and German were supplied by: Dr. John Biligmeir, Barbara Bolloch-Drake, Fred Boulle, Kathy Drake, Dr. Borimir Jordan, Angelika Pagel, Dr. Franchesca Sautman, and Dr. Gad Schnei.

Cheshire Books would like to thank the following individuals and firms for their help and patience in the editing and production of this book:

R. G. Beukers, copy editing and proofing
Deborah Wong, production assistant
Robert Cooney, paste-up
David Toze, photo work
Holly Lyman Antolini, final editing
Elinor Lindheimer, indexing
Lisanne Abraham

Paul Stanley, Arcata Book Group,
   Los Gatos, California
CBM Type, Sunnyvale, California
Field Engraving, Castro Valley, California
Graphicstat, Palo Alto, California
Photographics, Palo Alto, California
Frank's Type, Palo Alto, California

Text set in Times Roman, Mergenthaler VIP, by CBM Type, Sunnyvale, California. Printing and binding by Fairfield Graphics, Fairfield, Pennsylvania. Jacket printing by Hatcher Trade Press, San Carlos, California, with gold stamping by K. Hogan Co., Sunnyvale, California.

DATE			